INTRODUCTION TO
ASYMPTOTICS

A TREATMENT USING NONSTANDARD ANALYSIS

INTRODUCTION TO
ASYMPTOTICS

A TREATMENT USING NONSTANDARD ANALYSIS

D S Jones

Department of Mathematics, University of Dundee, Scotland

 **World Scientific**
Singapore • New Jersey • London • Hong Kong

Published by

World Scientific Publishing Co. Pte. Ltd.

P O Box 128, Farrer Road, Singapore 912805

USA office: Suite 1B, 1060 Main Street, River Edge, NJ 07661

UK office: 57 Shelton Street, Covent Garden, London WC2H 9HE

Library of Congress Cataloging-in-Publication Data
Jones, D. S. (Douglas Samuel)
  Introduction to asymptotics / D. S. Jones.
      p.   cm.
  Includes bibliographical references and index.
  ISBN 9810229151
  1. Asymptotic expansions.   I. Title.
QA295.J633   1966
519.2'4--dc20   96-36521
                                                          CIP

British Library Cataloguing-in-Publication Data
A catalogue record for this book is available from the British Library.

This book is printed on acid-free paper.

Printed in Singapore by Uto-Print

Dedicated to Muriel and Pat McPherson
the first to welcome and the last to say good-bye

Preface

There are several textbooks which deal comprehensively with the theory of asymptotic analysis and its applications (see references at the end of this book). No attempt to emulate them is made here since the purpose is to offer quick access to the principal facts in order that the reader can gain familiarity rapidly with this valuable tool. Some of the recent developments, not readily available outside the research literature, have been included. Also, the opportunity has been taken to illustrate, when convenient, how proofs of some results can be formulated by means of nonstandard analysis.

Although nonstandard analysis is an integral part of the presentation it is hoped that there is sufficient explanation for those who prefer a conventional approach to convert the arguments suitably. In any case there is an appendix giving a brief introduction to those ideas of nonstandard analysis needed to read the book.

After an opening chapter on general theory the topics of integrals, series, uniform asymptotics, hyperasymptotics and differential equations are dealt with in successive chapters. The aim has been to highlight the main methods deployed while setting forth their derivation in a way which is comprehensible to those who are new to the subject. Illustrative examples, both analytical and numerical, are placed at various stages in the chapters. In addition, there are exercises to enable readers to check their understanding of the techniques employed and, in some cases, to amplify what has been described in the text.

My thanks are due to my wife for her everlasting assistance and encouragement, and to Mrs. Ross who translated my manuscript into a beautiful typescript with her customary good humour and expertise.

Dundee, January 1996   D.S.Jones

Contents

Chapter 1
BASIC THEORY

1.1 Introduction

In early courses on infinite series students are introduced to the ideas of convergence and divergence. Usually most of the time is devoted to the properties of convergent series. Nevertheless, divergent series are of great practical importance and a considerable body of theory concerning them has been built up. To see why divergent series have a significant role to play a simple example will be discussed.

Let the function $f(x) = \int_x^\infty t^{-1}e^{x-t}dt$. We want to be able to estimate the numerical value of $f(x)$ when x is large. Repeated integration by parts furnishes

$$f(x) = \frac{1}{x} - \frac{1}{x^2} + \frac{2!}{x^3} - \cdots + \frac{(n-1)!(-)^{n-1}}{x^n} + n!(-)^n \int_x^\infty \frac{e^{x-t}}{t^{n+1}}dt. \quad (1.1.1)$$

If this process were continued indefinitely we should end up with a divergent series because the ratio of the mth and $(m-1)$th terms in Eq.(1.1.1) grows with m. To get a grasp of what happens as n increases write $R_n(x)$ for the last term of Eq.(1.1.1). Then, since $e^{x-t} \leq 1$ in the range of integration,

$$|R_n(x)| \leq n! \int_x^\infty \frac{dt}{t^{n+1}} = \frac{(n-1)!}{x^n}. \quad (1.1.2)$$

Although this is an inequality it suggests that the remainder $R_n(x)$ will become steadily larger as n grows. The increase will have to be balanced by the series because $f(x)$ is finite and so the infinite series can be expected to diverge. Yet all is not lost since Eq.(1.1.2) shows that, if n is held fixed, and x allowed to increase the remainder tends to zero. Thus, if n is fixed and the integral in Eq.(1.1.1) dropped, the series can give a good estimate of $f(x)$ when x is large enough.

For instance, if $n = 6$ and $x = 10$, the series gives the estimate 0.09152 for $f(10)$. According to Eq.(1.1.2) the error is less than 0.00012 indicating that the approximation is wrong, at worst, by about 1 in the fourth decimal place. In fact, the estimate is much better than that because the tabulated value of

$f(10)$ is 0.091563. For values of x larger than 10 one can expect the accuracy to improve on account of Eq.(1.1.2).

The reason why the first few terms of a divergent series give a good approximation in this case is that the remainder tends to zero as $x \to \infty$ with n fixed even though the remainder oscillates between $\pm\infty$ when $n \to \infty$ with x fixed. Of course, a good approximation could have been obtained also if the series had been convergent. The important point is that the remainder tends to zero as $x \to \infty$ with n fixed whether or not the infinite series is convergent.

Naturally, it is desirable to be able to derive estimates for values of x which are other than large. In this case a remainder which tends to zero as $x \to x_0$ (say) with n fixed is needed. Therefore a theory is required which covers remainders as x tends to an arbitrary value with n fixed. However, this is unnecessarily general. For $y = x - x_0$ makes $y \to 0$ when $x \to x_0$ and $y = 1/x$ makes $y \to 0$ when $x \to \infty$. Hence, it is sufficient to develop a theory which deals with behaviour near the origin and then use transformations to adapt the results for other points.

The methods used in this book are based on nonstandard analysis. For the benefit of the reader a brief account of the theory of nonstandard analysis is given in an appendix. Most of the theorems are accompanied by short proofs but a quick taste of the flavour can be achieved by skipping over them. A more comprehensive, but highly readable, description can be found in Robert (1988). Items in the appendix which are referred to subsequently will carry the prefix A e.g. Theorem A.2.1.

For those who wish to translate nonstandard statements into conventional ones the following rough guide may be of assistance

Nonstandard	*Replacement*
Infinitesimal ν	$o(1)$
Limited number	Medium-sized number
Unlimited number	Large number
$f(x)$ is limited	$f(x) = O(1)$.

Roughly speaking, the value of $f(x)$ when x is infinitesimal is not far off $\lim_{x \to 0} f(x)$ and when x is unlimited positive the value is pretty much the same as $\lim_{x \to \infty} f(x)$ provided that the limits exist.

Asymptotic theory covers much more than is dealt with in this book and does not have to be approached through nonstandard analysis. For more comprehensive treatments see Bleistein & Handelsman (1975), Olver (1974), Wong (1989) who employ conventional techniques, Jones (1982), Lighthill (1958) who deploy generalised functions and van den Berg (1987) who works with nonstandard analysis.

1.2　Asymptotic sequences

The series in the example discussed in Section 1.1 was in powers of $1/x$. A general theory needs to cope with expansions in more general functions of x. Not all functions are suitable for asymptotic estimates so this section is concerned with indicating the sort of function which can be appropriate.

As explained in the preceding section we are going to consider behaviour near the origin and rely on transformations to convert the theorems to apply at other points. Therefore, it will be assumed that functions of a complex variable z are being handled in some neighbourhood of the origin. In specific cases z may be restricted to the real line or have its phase limited in some way. At this stage it is inconvenient to make constant reference to any such limitations. So a phrase such as 'for all z with $|\mathrm{ph}\,z| < \pi/4$' will be abbreviated to 'for all z' with any restrictions understood implicitly. Admittedly, there is a certain loss of strictness, but it should occasion no difficulty so long as any restrictions are mentioned explicitly when that is vital.

Since z is going to be near the origin it will be convenient often to take it as infinitesimal. A non-zero infinitesimal is very small in magnitude; its reciprocal is an unlimited number which is very large but finite. Proving theorems for infinitesimals may constrain their applicability but, normally, they carry over to larger values, particularly when reliable error estimates are available.

As to notation the symbol $\simeq$ between quantities signifies that they differ by an infinitesimal e.g. $x \simeq y$ is another way of saying that $x - y$ is infinitesimal. The Russian и will denote a generic infinitesimal which need not be the same wherever it occurs. Thus

$$2и = и, \quad \ln(1 + и) = и$$

are legitimate statements. So is $bи = и$ when b is limited (see appendix) but may not be when b is unlimited. When a specific invariable infinitesimal is required a suffix will be attached to и.

Most of the theory will be developed in terms of standard functions. There are two reasons for this. The first is that uniquely defined conventional functions are standard so that the theory should be wide enough in scope. The second reason is that statements involving standard quantities can be extended by the transfer rule (see appendix).

There is an algorithm which converts a nonstandard statement into a classical one (Nelson 1977). In essence, the argument goes as follows. Suppose that the standard function f is such that $f(\epsilon) \simeq 0$ for all infinitesimal ϵ. Then, for any standard $\eta > 0$, $|f(\epsilon)| < \eta$ for all infinitesimal ϵ. Choose some infinitesimal ϵ_0 and put $\delta = \epsilon_0$. We have, for every standard $\eta > 0$, there is $\delta > 0$ such that $|f(\epsilon)| < \eta$ for all ϵ with $|\epsilon| < \delta$. This statement contains only the free parameters f and η which are standard. Hence the transfer rule validates the classical statement: for all $\eta > 0$ there is $\delta > 0$ such that $|f(x)| < \eta$ for all $|x| < \delta$.

Definition 1.2.1 *The sequence of standard functions* $\{\varphi_n(z)\}$ *is said to be an asymptotic sequence as* $z \to 0$ *when, for each* n, $\varphi_{n+1}(z) = \varkappa\varphi_n(z)$ *for all* $z \simeq 0$.

We stress that $\varkappa$ is generic so that it can change with n.

Notice that the condition has to be verified only for infinitesimal z. Clearly, then, $\{z^n\}$ is an asymptotic sequence. When z is real $\{z^{\lambda_n}\}$ is an asymptotic sequence for $\mathcal{R}(\lambda_{n+1}) > \mathcal{R}(\lambda_n)$ (every n); for z complex it may be necessary to have a branch line along the negative real axis and then $\{z^{\lambda_n}\}$ is asymptotic for $|\mathrm{ph}\, z| < \pi - \delta$ with $\delta > 0$. Under the same conditions $\{\varphi(z)z^{\lambda_n}\}$ with φ non-zero is an asymptotic sequence because the additional factor disappears from the condition of the definition.

Sometimes circumstances permit a limitation on the phase of z to be surmounted provided that any variation is carefully tracked. In effect, z is allowed to move on a Riemann surface with its phase changing continuously even when a branch line is crossed. In that case z^{λ_n} is understood to mean $|z|^{\lambda_n} \exp(i\lambda_n \,\mathrm{ph}\, z)$. Suppose that the phase of z is increased continuously until it has grown by 2π. Then z^{λ_n} is changed to $|z|^{\lambda_n} \exp\{i\lambda_n(\mathrm{ph}\, z + 2\pi)\}$ which is different from its previous value unless λ_n is an integer. This explains why the phase of z has to be watched carefully. However, $\{z^{\lambda_n}\}$ can be regarded now as asymptotic over a larger region than in the preceding paragraph.

A somewhat more complicated example is $\{z^{\lambda_n} \exp(-n/z)\}$. Here there need be no restrictions on λ_n for an asymptotic sequence if $|\mathrm{ph}\, z| < \frac{1}{2}\pi - \delta$ but, for $|\mathrm{ph}\, z| \le \frac{1}{2}\pi$, $\mathcal{R}(\lambda_{n+1}) > \mathcal{R}(\lambda_n)$ would have to be imposed.

Remark that Definition 1.2.1 implies that $\varphi_{n+m}(z) = \varkappa^m\varphi_n(z)$ for any positive integer m. Hence *any subsequence of an asymptotic sequence is asymptotic.*

Again, $\{|\varphi_n(z)|\}$ is an asymptotic sequence when $\{\varphi_n(z)\}$ is and, indeed, so is $\{|\varphi_n(z)|^\alpha\}$ with $\alpha > 0$. By inserting the condition of Definition 1.2.1 into each factor of $\varphi_n\psi_n$ we see that $\{\varphi_n(z)\psi_n(z)\}$ is an asymptotic sequence when both $\{\varphi_n\}$ and $\{\psi_n\}$ are asymptotic.

Let ν be a generic limited number, not necessarily the same at each occurrence. Then we have

Theorem 1.2.1 *If, for every* n, $\varphi_n(z) = \nu\psi_n(z)$ *and* $\psi_n(z) = \nu\varphi_n(z)$ *then, when* $\{\varphi_n(z)\}$ *is asymptotic, so is* $\{\psi_n(z)\}$.

Proof. $\psi_{n+1}(z) = \nu\varphi_{n+1}(z) = \nu\varkappa\varphi_n(z) = \nu\varkappa\nu\psi_n(z) = \varkappa\psi_n(z)$ since $\nu\varkappa \simeq 0$ when ν is limited. ∎

Let $\varkappa_0$ be a fixed infinitesimal and let a_{ni} be standard positive constants with $a_{n+1,i} \le a_{ni}$ for $n = 1, 2, \ldots, i = 0, 1, \ldots$.

Theorem 1.2.2 *If* $\varphi_{n+1}(z) = \varkappa_0\varphi_n(z)$ *for every* n *and* $\sum_{i=0}^{\infty} a_{1i}|\varkappa_0|^i$ *is limited*

then $\{\psi_n(z)\}$ is an asymptotic sequence where

$$\psi_n(z) = \sum_{i=0}^{\infty} a_{ni} \left| \varphi_{n+i}(z) \right|.$$

Proof. $\sum_{i=0}^{\infty} a_{ni} \left| \varkappa_0 \right|^i \leq \sum_{i=0}^{\infty} a_{n-1,i} \left| \varkappa_0 \right|^i \leq \cdots \leq \sum_{i=0}^{\infty} a_{1i} \left| \varkappa_0 \right|^i$ shows that all series of this type are limited by hypothesis. Therefore

$$\psi_n(z) = \left| \varphi_n(z) \right| \sum_{i=0}^{\infty} a_{ni} \left| \varkappa_0 \right|^i$$

and

$$\psi_{n+1}(z) \leq \left| \varphi_{n+1}(z) \right| \sum_{i=0}^{\infty} a_{ni} \left| \varkappa_0 \right|^i$$

whence the theorem follows. ∎

A corresponding theorem for integrals instead of sums could be devised but would be superfluous to our purposes. Observe, however, that the derivative of an asymptotic sequence may not be asymptotic. A counterexample is furnished by $\{z^n [\cos(1/z^n) + 2]\}$ when z is real.

1.3   Asymptotic expansions

Asymptotic sequences have been introduced in order that functions can be represented by expansions of them. We would like to write

$$f(z) = c_1 \varphi_1(z) + c_2 \varphi_2(z) + \cdots$$

with $c_1, c_2, \ldots$ constants. However, the series when infinite might diverge and this possibility has to be allowed for.

Definition 1.3.1 *Let $\{\varphi_n(z)\}$ be an asymptotic sequence and c_m a standard constant for standard m. Let*

$$f(z) = \sum_{m=1}^{n} c_m \varphi_m(z) + R_n(z),$$

f being standard. If, for $z \simeq 0$,

$$R_n(z) = (c_{n+1} + \varkappa) \varphi_{n+1}(z)$$

for all standard n, $\sum_{m=1}^{\infty} c_m \varphi_m(z)$ is said to be an asymptotic expansion for f as $z \to 0$ and we write

$$f(z) \sim \sum_{m=1}^{\infty} c_m \varphi_m(z).$$

It is understood that conditions on the phase of z may have to be imposed and that the infinite series may not converge.

Theorem 1.3.1 *The Maclaurin series of a function with derivatives continuous at the origin is asymptotic.*

Proof. The customary series is

$$f(z) = \sum_{m=0}^{n} f^{(m)}(0)\frac{z^m}{m!} + f^{(n+1)}(\theta z)\frac{z^n}{n!}$$

with $0 < \theta < 1$. According to Section A.4 $f^{(n)}$ is standard when f is. Also, at the standard point 0, $f^{(n+1)}$ is continuous by assumption and so S-continuous on account of Theorem A.3.1. Hence, for $z \simeq 0$, $f^{(n+1)}(\theta z) \simeq f^{(n+1)}(0)$ and the theorem is proved. ∎

The formula of Definition 1.3.1 can be written as

$$\{f(z) - \sum_{m=1}^{n} c_m\varphi_m(z)\}/\varphi_{n+1}(z) = c_{n+1} + и \qquad (1.3.1)$$

from which follows, for all standard n,

$$c_{n+1} = \mathrm{st}[\{f(z) - \sum_{m=1}^{n} c_m\varphi_m(z)\}/\varphi_{n+1}(z)] \qquad (1.3.2)$$

for $z \simeq 0$, where st means standard part. Take the c_{n+1} over to the left-hand side in Eq.(1.3.1) and apply Corollary A.2.2. Then

$$c_{n+1} = \lim_{z \to 0}\{f(z) - \sum_{m=1}^{n} c_m\varphi_m(z)\}/\varphi_{n+1}(z) \qquad (1.3.3)$$

for all standard n. The transfer rule confirms Eq.(1.3.3) for all $n \in \mathbf{N}$.

Either Eq.(1.3.2) or Eq.(1.3.3) demonstrates that the coefficients in the asymptotic expansion of $f(z)$ are determined uniquely by f and the asymptotic sequence. In other words, if $f(z) \sim \sum_{m=1}^{\infty} c_m\varphi_m(z)$ and $f(z) \sim \sum_{m=1}^{\infty} C_m\varphi_m(z)$, it is necessary that $c_m = C_m$ for all m. Different expansions can be obtained for the same function with different asymptotic sequences. For example

$$(1+z)^{-1} \sim \sum_{m=0}^{\infty} (-z)^m, \qquad (1.3.4)$$

$$(1+z)^{-1} \sim \sum_{m=0}^{\infty} (1-z)z^{2m}.$$

Although f determines an asymptotic expansion uniquely the converse is not true in general. Different functions can have the same asymptotic expansion. For example,

$$\{1 + \exp(-1/z)\}(1+z)^{-1}$$

has the same asymptotic expansion as Eq.(1.3.4) when z is positive real and $\{z^n\}$ is the asymptotic sequence. The exponential is totally negligible when the coefficients are calculated by Eq.(1.3.3).

Given an expansion in the $\varphi_m(z)$ it is always possible to construct a function for which it is asymptotic. Since it is only one of many possibilities it is not clear that such a construction will serve any practical purpose.

1.4   Operations on asymptotic expansions

Asymptotic expansions can be combined always in a linear fashion.

Theorem 1.4.1 *If $f(z) \sim \sum_{m=1}^{\infty} c_m \varphi_m(z)$ and $g(z) \sim \sum_{m=1}^{\infty} d_m \varphi_m(z)$ then*

$$af(z) + bg(z) \sim \sum_{m=1}^{\infty} (ac_m + bd_m)\varphi_m(z)$$

for a and b standard constants.

Proof. The remainder after n terms is

$$\{a(c_{n+1} + \varkappa) + b(d_{n+1} + \varkappa)\}\varphi_{n+1}(z).$$

Since a and b are limited $a\varkappa + b\varkappa$ is infinitesimal and the conditions of Definition 1.3.1 are met. ■

When $f(z, w)$ includes the extra variable w the coefficients in the asymptotic expansion will depend on w. In general, the infinitesimal $\varkappa$ in the remainder $R_n(z)$ of Definition 1.3.1 will depend on w also. Suppose that there is an infinitesimal $\varkappa_n$ independent of w such that $|\varkappa| \leq |\varkappa_n|$. Then we have

Theorem 1.4.2 *Under the given condition if $h(w)$ is absolutely integrable and $\int_{w_1}^{w_2} h(w)f(z, w)dw$, $C_n = \int_{w_1}^{w_2} h(w)c_n(w)dw$ exist then*

$$\int_{w_1}^{w_2} h(w)f(z, w)dw \sim \sum_{n=1}^{\infty} C_n \varphi_n(z).$$

Proof. Take w_1, w_2 standard first. Then

$$\int_{w_1}^{w_2} h(w)R_n(z, w)dw = \varphi_{n+1}(z)\{C_{n+1} + \int_{w_1}^{w_2} \varkappa h(w)dw\}$$

and the last integral is bounded by a limited multiple of $|\varkappa_n|$ which makes it infinitesimal. Thus, the theorem is proved for standard w_1, w_2 and the transfer rule extends it to more general w_1, w_2. ■

An analogous theorem for the derivative is precluded by the derivative of an asymptotic sequence not being an asymptotic sequence in general. In some specific cases this objection can be circumvented and a theorem derived but details are left to later (see Section 1.5).

Multiplication of asymptotic expansions is another operation which is not straightforward. The reason is that in $(\sum_n c_n\varphi_n(z))(\sum_m d_m\varphi_m(z))$ all products of the form $\varphi_n(z)\varphi_m(z)$ arise and it may not be possible to arrange them in an order which supplies an asymptotic sequence. However, it is sometimes possible to find asymptotic expansions of the products $\varphi_n(z)\varphi_m(z)$ in terms of another asymptotic sequence and then some progress can be made.

Let $\{\varphi_n(z)\}, \{\psi_n\}, \{\theta_n\}$ be asymptotic sequences such that $\varphi_1\psi_1 = \nu\theta_k$ for any k where the limited ν is generic. The definition of an asymptotic sequence warrants $\varphi_n\psi_m = \nu\theta_k$ for any n, m, k. Suppose that

$$\varphi_n\psi_m \sim \sum_{k=1}^{\infty} c_{nmk}\theta_k. \tag{1.4.1}$$

Theorem 1.4.3 *Let* $f(z) \sim \sum_{n=1}^{\infty} c_n\varphi_n(z)$, $g(z) \sim \sum_{m=1}^{\infty} d_m\psi_m(z)$ *with* $\{\varphi_n\}$, $\{\psi_n\}, \{\theta_n\}$ *having the properties just specified. Then, for limited* N *and* M,

$$f(z)g(z) \sim \sum_{k=1}^{\infty} C_k\theta_k(z)$$

where

$$C_k = \sum_{n=1}^{N}\sum_{m=1}^{M} c_n d_m c_{nmk}.$$

Proof. Evidently

$$
\begin{aligned}
f(z)g(z) &= \{\sum_{n=1}^{N} c_n\varphi_n(z) + \nu\varphi_N(z)\}\{\sum_{m=1}^{M} d_m\psi_m(z) + \nu\psi_M(z)\} \\
&= \sum_{n=1}^{N}\sum_{m=1}^{M} c_n d_m\varphi_n(z)\psi_m(z) + \nu\varphi_N(z)\sum_{m=1}^{M} d_m\psi_m(z) \\
&\quad + \nu\psi_M(z)\sum_{n=1}^{N} c_n\varphi_n(z) + \nu\varphi_N(z)\psi_M(z) \\
&= \sum_{k=1}^{K} C_k\theta_k + \nu\theta_K\sum_{n=1}^{N}\sum_{m=1}^{M} c_n d_m + \nu\nu\theta_K\sum_{m=1}^{M} d_m \\
&\quad + \nu\nu\theta_K\sum_{n=1}^{N} c_n + \nu\theta_K
\end{aligned}
$$

for any K. The sums of a finite number of standard terms are limited and so the theorem is proved. ∎

It is clear that the above analysis remains valid for infinite N and M provided that the various series stay limited. Accordingly we can state

Corollary 1.4.3 *If* $\sum_{n=1}^{\infty}\sum_{m=1}^{\infty} c_n d_m c_{nmk}$ *is standard and* $\sum_{n=1}^{\infty} c_n, \sum_{m=1}^{\infty} d_m$ *are limited*

$$f(z)g(z) \sim \sum_{k=1}^{\infty} C_k\theta_k(z)$$

where

$$C_k = \sum_{n=1}^{\infty} \sum_{m=1}^{\infty} c_n d_m c_{nmk}.$$

Observe that in both Theorem 1.4.3 and Corollary 1.4.3 the stated expansion can be obtained by first multiplying together formally the asymptotic expansions of f and g followed by substitution from Eq.(1.4.1).

It may happen that $\{\varphi_n(z)\}$ is such that we can choose to have $\psi_n = \varphi_n$ and $\theta_n = \varphi_n$. Then further multiplication can be undertaken. To keep things simple only the analogue of Theorem 1.4.3 will be given—that for Corollary 1.4.3 can be inferred by putting enough conditions on the various series involved.

Corollary 1.4.3a *If $\varphi_1(z)$ is limited and*

$$\varphi_n(z)\varphi_m(z) \sim \sum_{k=1}^{\infty} c_{nmk}\varphi_k(z)$$

then

$$\{f(z)\}^2 \sim \sum_{k=1}^{\infty} C_k \varphi_k(z)$$

where

$$C_k = \sum_{n=1}^{N} \sum_{m=1}^{N} c_n c_m c_{nmk}.$$

More generally, $\{f(z)\}^p$ with p a standard integer has an asymptotic expansion in terms of $\{\varphi_n(z)\}$, the coefficients being determined by formal substitution.
Proof. The formula for $\{f(z)\}^2$ is a direct consequence of Theorem 1.4.3 with the conditions imposed on $\{\varphi_n(z)\}$ and the coefficients are obtained by formal substitution. Since $\{f(z)\}^2$ and $f(z)$ both have asymptotic expansions in $\{\varphi_n(z)\}$ they can be multiplied to give $\{f(z)\}^3$ with coefficients determined by formal substitution. The process can be continued to reach $\{f(z)\}^p$ when p is a standard integer. ∎

By tightening somewhat the restriction on $\varphi_1(z)$ these results can be extended in a valuable way. Let $\varphi_1(z)$ be infinitesimal for infinitesimal z and suppose that $\{\varphi_1(z)\}^n = \nu\varphi_n(z)$ for limited ν. Since $\varphi_1(z)$ is the dominant term in $f(z)$ it is clear that the dominant term in $\{f(z)\}^p$ is $\{\varphi_1(z)\}^p$. It is feasible now to assert
Theorem 1.4.4 *If $F(w) \sim \sum_{n=1}^{\infty} d_n w^n$ and $f(z) \sim \sum_{m=1}^{\infty} c_m \varphi_m(z)$ then*

$$F(f(z)) \sim \sum_{n=1}^{\infty} C_n \varphi_n(z)$$

where the coefficients C_n are obtained by formal substitution.
Proof. With N standard

$$F(f(z)) = \sum_{n=1}^{N} d_n \{f(z)\}^n + \mu \{f(z)\}^N$$

so that

$$F(f(z)) = \sum_{n=1}^{N} d_n \{ \sum_{m=1}^{N} c_{nm} \varphi_m(z) + и \varphi_N(z) \} + и \{ f(z) \}^N.$$

Since $\{f(z)\}^N = \nu \varphi_N(z)$ by the assumption on $\varphi_1(z)$ the desired expansion is confirmed. ∎

An application of this theorem shows that it is possible to divide by an asymptotic expansion. For, when $c \neq 0$,

$$1/(c+w) \sim \sum_{m=0}(-w)^m/c^{m+1}.$$

Indeed, the series is actually convergent for $|w| < |c|$. Write the right-hand side as $1/c + F(w)$ and apply Theorem 1.4.4.

Corollary 1.4.4 *If $c \neq 0$, $1/\{c+f(z)\}$ has an asymptotic expansion in $\{\varphi_n(z)\}$ when $\varphi_1(z)$ is subject to the same conditions as in Theorem 1.4.4.*

An asymptotic series can be inverted in certain circumstances. Suppose that

$$w(z) \sim 1/z + c_0 + \sum_{m=1}^{\infty} c_m z^m$$

and it is desired to find z as a function of w. It will be assumed that conditions are such that to a given w there corresponds one, and only one, z. Evidently, w is large and, to a first approximation, $z = 1/w$. In fact, w must be unlimited when z is infinitesimal. The next approximation satisfies $zw = 1 + (c_0 + и)z$ which shows that $z = 1/w + (c_0 + и)/w^2$. Bringing in c_1 improves the approximation and the introduction of one coefficient at a time leads to

$$z = \frac{1}{w} + \frac{c_0}{w^2} + \frac{c_1}{w^3} + \frac{d_2}{w^4} + \cdots + \frac{d_{n-2} + и}{w^n}$$

which provides z with an asymptotic expansion in the infinitesimal $1/w$.

1.5   Power series

A particular example of an asymptotic sequence is $\{z^n\}$. It occurs so frequently in asymptotic expansions that it is worth examining what happens when the foregoing theory is applied.

For the sequence $\{z^n\}$ a typical asymptotic expansion is

$$f(z) \sim c_0 + c_1 z + c_2 z^2 + \cdots. \tag{1.5.1}$$

From Theorem 1.4.1 it is evident that two expansions of the type of Eq.(1.5.1) can be multiplied by constants and added. They can be multiplied together also because the conditions of Theorem 1.4.3 are satisfied. So long as $c_0 \neq 0$ division is possible as well (Corollary 1.4.4). Substitution in polynomials is permissible by Corollary 1.4.3a. On account of Theorem 1.4.4 they can replace w in the asymptotic $\sum d_n w^n$ so long as $c_0 = 0$. Inversion of Eq.(1.5.1) was discussed at the end of the last section. Theorem 1.4.2 is relevant to integration with respect to a parameter.

In addition, Eq.(1.5.1) may be integrated with respect to z term-by-term. For

$$
\begin{aligned}
\int_0^z f(t)dt &= \int_0^z (\sum_{n=0}^N c_n t^n + \varkappa t^N)dt \\
&= \sum_{n=0}^N c_n z^{n+1}/(n+1) + \varkappa z^{N+1}.
\end{aligned}
$$
(1.5.2)

If it is known that $f'(z) \sim \sum_{m=0}^\infty d_m z^m$ replace f by f' in Eq.(1.5.2). Then

$$
f(z) - f(0) \sim \sum_{m=0}^\infty d_m z^{m+1}/(m+1).
$$

Comparison with Eq.(1.5.1) yields $c_0 = f(0)$, $d_m = (m+1)c_{m+1}$ on account of the uniqueness of the coefficients in an asymptotic expansion. This result is sufficiently important to state as a theorem.

Theorem 1.5.1 *If the $f(z)$ of Eq.(1.5.1) is differentiable and $f'(z)$ possesses an asymptotic expansion in $\{z^n\}$ then*

$$
f'(z) \sim \sum_{n=1}^\infty n c_n z^{n-1}.
$$

When the expansion of Eq.(1.5.1) refers to a function regular for $|z| \leq r$, r standard, over a sufficient range of ph z a more general theorem is available. Let the region of regularity include $\alpha \leq$ ph $z \leq \beta$ with α, β standard and $\alpha < \beta$. Then f is a regular function of z in the sectoral region so prescribed (Fig. 1.5.1). The existence of $f'(z)$ at an interior point is then assured. Pick a point z in the region so that a circle C of radius $\epsilon |z|$ and centre z can be drawn entirely within the sector. By Cauchy's theorem

$$
f'(z) = \frac{1}{2\pi i} \int_C \frac{f(w)}{(w-z)^2} dw.
$$

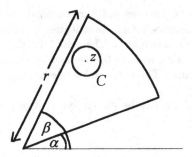

Figure 1.5.1 Sector of regularity

On the circle put $w = z(1 + \epsilon e^{i\theta})$; then

$$f'(z) = \frac{1}{2\pi z\epsilon} \int_0^{2\pi} e^{-i\theta} f\{z(1 + \epsilon e^{i\theta})\} d\theta.$$

If the infinitesimal in the remainder of Definition 1.3.1 is independent of the phase of z (the expansion is said then to be *uniform* in angle) Eq.(1.5.1) can be substituted into the integral as $z \to 0$. Since

$$\int_0^{2\pi} e^{-i\theta}(1 + \epsilon e^{i\theta})^n d\theta = n\epsilon$$

the same formula as in Theorem 1.5.1 is recovered. Obviously, the expansion will be uniform in angle but in a smaller sector, say $\alpha + \delta \le \mathrm{ph}\, z \le \beta - \delta$ with $\delta > 0$, in order to allow room for the circle C to be drawn.

There is a similar Cauchy integral for the mth derivative $f^{(m)}(z)$ so that the same route may be traversed to find an asymptotic expansion for it.

Theorem 1.5.2 *If $f(z)$ is regular in $\alpha \le \mathrm{ph}\, z \le \beta$ and Eq.(1.5.1) is uniform in angle then $f^{(m)}(z)$ has an asymptotic expansion in $\alpha + \delta \le \mathrm{ph}\, z \le \beta - \delta$ obtained by taking derivatives of Eq.(1.5.1) term-by-term and it is uniform in angle.*

1.6   A generalisation

Not every approximating series fits the structure of Definition 1.3.1. For instance, the series

$$\sum_{n=1}^{\infty} z^n \sin(n\pi - 1/z)$$

might be met. Although this series converges absolutely as $z \to 0$ it is not an asymptotic expansion in the sense of Definition 1.3.1 because the ratio of successive terms is not bounded. To include such series in a definition it has been suggested that

$$f(z) = \sum_{m=1}^{n-1} f_m(z) + \nu \varphi_n(z)$$

for every n with $\{\varphi_n\}$ an asymptotic sequence should be accepted as defining an asymptotic expansion. However, with such generality, most of the theorems developed above become inapplicable and an adequate theory is lacking. Therefore, when series outside the scope of Definition 1.3.1 occur, they will have to be dealt with on an individual ad hoc basis.

1.7   An illustration

One way of establishing an asymptotic expansion by approximating an integral and then applying Eq.(1.3.2) will be illustrated now.

Theorem 1.7.1 *Let ϵ be an infinitesimal with $|\mathrm{ph}\,\epsilon| \le \frac{1}{2}\pi$. Then, for any positive $\lambda \in \mathbf{R}$,*

$$\int_0^\infty \frac{e^{-t}}{(1+\epsilon t)^\lambda} dt = \frac{1+\varkappa}{1+\lambda\epsilon}$$

where $(1+\epsilon t)^\lambda = 1$ when $t = 0$.

Proof. First let $\lambda\,|\epsilon|$ be limited. For limited t, $t\epsilon$ is infinitesimal and

$$\lambda \ln(1+\epsilon t) = \lambda\epsilon t(1+\varkappa) = \lambda t\epsilon + \varkappa$$

since $\lambda t\epsilon$ is limited. Therefore

$$(1+t\epsilon)^{-\lambda} \simeq e^{-\lambda t\epsilon}$$

for limited t. Theorem A.5.5 can be applied providing that a suitable bounding function for all t is available. Now $|1 + \lambda\epsilon| > 1$ and $|1 + t\epsilon| > 1$ so that both $\left|e^{-t}(1+t\epsilon)^{-\lambda}\right|$ and $\left|e^{-t-\lambda t\epsilon}\right|$ are bounded by e^{-t} which is integrable. Hence, by Theorem A.5.5,

$$\int_0^\infty \frac{e^{-t}}{(1+t\epsilon)^\lambda} dt \simeq \int_0^\infty e^{-t-\lambda t\epsilon} dt = \frac{1}{1+\lambda\epsilon}.$$

Thus the theorem has been demonstrated when $\lambda\epsilon$ is limited.

Consider

$$(1+\lambda\epsilon) \int_0^\infty \frac{e^{-t}}{(1+t\epsilon)^\lambda} dt - 1$$

which has been shown to be infinitesimal when $\lambda\epsilon$ is limited. Keep ϵ fixed and treat $\lambda\,|\epsilon|$ as a continuous variable x. Since the function is infinitesimal for all limited x Robinson's Continuous Lemma may be quoted to verify that it is infinitesimal for x up to some unlimited value. In other words, the theorem holds for $\lambda \le \lambda_0$ where $\lambda_0 \epsilon$ is unlimited. In fact, when $\lambda = \lambda_0$ the 1 in the denominator could be dropped as insignificant but is retained so that one formula is relevant for all λ.

To extend the formula beyond λ_0 choose λ so that $\lambda_0 - 1 \le \lambda \le \lambda_0$ and write $\varkappa_0$ for the infinitesimal $1/\lambda |\epsilon|$. Then, for any positive integer p,

$$\int_0^\infty \frac{e^{-t}}{(1+t\epsilon)^{\lambda+p}} dt = \frac{1}{(\lambda+p-1)\epsilon} \left\{ 1 - \int_0^\infty \frac{e^{-t} dt}{(1+t\epsilon)^{\lambda+p-1}} \right\} \qquad (1.7.1)$$

by integration by parts. Repetition of the process supplies

$$\int_0^\infty \frac{e^{-t}}{(1+t\epsilon)^{\lambda+p}} dt$$

$$= \frac{1}{(\lambda+p-1)\epsilon} \left[1 - \frac{1}{(\lambda+p-2)\epsilon} + \frac{1}{(\lambda+p-3)(\lambda+p-2)\epsilon^2} \right.$$
$$\left. - \cdots + \frac{(-)^{p-1}}{\lambda(\lambda+1)\dots(\lambda+p-2)\epsilon^{p-1}} \left\{ 1 - \int_0^\infty \frac{e^{-t}}{(1+t\epsilon)^\lambda} dt \right\} \right].$$

What has been proved already shows that the quantity in [ ] does not exceed in modulus

$$1 + \varkappa_0 + \varkappa_0^2 + \cdots + \varkappa_0^p = 1 + \varkappa$$

for any p. Since $(\lambda+p-1)\epsilon = (1+\varkappa)\{1 + (\lambda+p)\epsilon\}$ the theorem has been proved without restriction on λ. ∎

Next the asymptotic expansion of

$$\int_0^\infty \frac{e^{-t}}{1+t\epsilon} dt$$

in terms of the asymptotic sequence $\{\epsilon^n\}$ is discussed. There are several ways of obtaining the expansion but the method chosen here illustrates how Eq.(1.3.2) can be applied. According to Eq.(1.3.2)

$$c_0 = \text{st} \int_0^\infty \frac{e^{-t}}{1+t\epsilon} dt = \text{st} \frac{1+\varkappa}{1+\epsilon} = 1$$

by Theorem 1.7.1. Further

$$c_1 = \text{st} \left(\int_0^\infty \frac{e^{-t}}{1+t\epsilon} dt - 1 \right) \frac{1}{\epsilon} = \text{st} - \int_0^\infty \frac{e^{-t}}{(1+t\epsilon)^2} dt = -1$$

from Eq.(1.7.1) and Theorem 1.7.1. Again

$$c_2 = \text{st} \left(-\int_0^\infty \frac{e^{-t}}{(1+t\epsilon)^2} dt + \epsilon \right) \frac{1}{\epsilon^2} = \text{st} \, 2 \int_0^\infty \frac{e^{-t}}{(1+t\epsilon)^3} dt = 2$$

and generally $c_n = n!(-)^n$ for standard n. The same procedure cannot be adopted when n is unlimited because (a) Eq.(1.3.2) has been stated only for standard n and (b) c_n becomes unlimited. However, $\{n!(-)^n\}$ is a standard sequence which agrees with c_n when n is standard and so can be used to extend the range of c_n (see also the remark after Corollary A.2.2a). Thus, we have

Theorem 1.7.2 *For infinitesimal ϵ with $|\mathrm{ph}\,\epsilon| \leq \frac{1}{2}\pi$*

$$\int_0^\infty \frac{e^{-t}}{1+t\epsilon}dt \sim \sum_{n=0}^\infty n!(-\epsilon)^n.$$

Exercises on Chapter 1

1. Do $\ln z$ and $\sin(1/z)$ possess asymptotic expansions based on the sequence $\{z^n\}$ according to Definition 1.3.1?

2. Let $f(z) \sim \sum_{m=0}^\infty c_m z^m$ with $c_0 \neq 0$ and $1/f(z) \sim \sum_{m=0}^\infty d_m z^m$. Use the fact that $f(z)/f(z) = 1$ to prove that $d_0 = 1/c_0$ and

$$d_n c_0 = -(d_{n-1}c_1 + d_{n-2}c_2 + \cdots + d_0 c_n)$$

for $n = 1, 2, \ldots$. Show that $d_1 = -c_1/c_0^2$ and $d_2 = (c_1^2 - c_0 c_2)/c_0^3$.

3. If $f(z) \sim 1 + \sum_{m=1}^\infty c_m z^m$ and $\ln f(z) \sim \sum_{m=1}^\infty d_m z^m$ show that $d_1 = c_1$ and $nd_n = nc_n - (n-1)d_{n-1}c_1 - (n-2)d_{n-2}c_2 - \cdots - d_1 c_{n-1}$ for $n = 2, 3, \ldots$.

4. If z is real and $\{\varphi_n(z)\}$ is an asymptotic sequence of positive continuous functions show that, when f is continuous and $f(z) \sim \sum_{n=1}^\infty c_n \varphi_n(z)$,

$$\int_0^z f(t)dt \sim \sum_{n=1}^\infty c_n \int_0^z \varphi_n(t)dt.$$

5. Find an asymptotic expansion of $\int_0^\infty \frac{e^{-t}}{(1+\epsilon t)^3}dt$ by the method of Section 1.7.

6. If λ is limited show that Theorem 1.7.1 holds for $|\mathrm{ph}\,\epsilon| \leq \pi - \delta$ with standard $\delta > 0$.

7. Each $\varphi_n(z)$ in the asymptotic sequence $\{\varphi_n(z)\}$ is absolutely integrable. Show that $\{\int_0^z |\varphi_n(t)|\,dt\}$ is an asymptotic sequence.

Chapter 2
INTEGRALS

2.1   Stirling's formula

The first type of integral to be considered for asymptotic behaviour includes a representation of the factorial function. Therefore, it supplies Stirling's formula for the factorial as a special case.

Theorem 2.1.1 *Let ϵ be an infinitesimal with phase α. Let $f(t)$ be S-continuous at $t = \mu\epsilon$ and such that $|f(re^{i\alpha})| \le Ae^{\gamma r}$ for all $r \ge 0$ with A and γ standard constants. Then, for $\mu > -1$ and limited $\mu\epsilon$,*

$$\int_0^\infty e^{-t} t^\mu f(\epsilon t)dt = \mu! f(\mu\epsilon)(1 + \varkappa).$$

Proof. The existence of the integral is warranted by the bound on f and $\gamma\epsilon$ being infinitesimal so that the integrand is certainly bounded by $At^\mu e^{-t/2}$.

Consider first the case when μ is limited so that $\mu\epsilon$ is infinitesimal. When t is limited, $t\epsilon$ is infinitesimal and $f(\epsilon t) \simeq f(\mu\epsilon)$ by S-continuity. The integrand is bounded for all t as explained already. It follows from Theorem A.5.5 that the formula stated holds.

Next, let μ be unlimited but such that $\mu\epsilon$ is limited. Make the substitution $t = \mu u$, so that the integral becomes

$$\mu^{\mu+1} \int_0^\infty \exp\{\mu(\ln u - u)\} f(\mu\epsilon u)du.$$

The argument of the exponential is always negative. As u increases from zero it rises from $-\infty$ to a maximum of $-\mu$ when $u = 1$ and thereafter decreases to $-\infty$. In view of the largeness of μ this suggests that an interval around $u = 1$ will be the most important. So, the range of integration will be split into three portions with the central piece containing a small neighbourhood of $u = 1$.

Let $\theta = 1/\mu^{1/4}$; accordingly θ is infinitesimal. For $u \ge 1 + \theta$,

$$\ln u \le \ln(1 + \theta) - 1 + u/(1 + \theta). \tag{2.1.1}$$

17

Also θ is much larger than $\gamma\,|\epsilon|$ because $\mu\theta$ is unlimited whereas $\mu\gamma\,|\epsilon|$ is limited. Hence

$$\left| \mu^{\mu+1} \int_{1+\theta}^{\infty} \exp\{\mu(\ln u - u)\} f(\mu\epsilon u) du \right|$$

$$\leq A\mu^{\mu+1} \exp[\mu\ln(1+\theta) - \mu + \mu\{(1+\theta)\gamma\,|\epsilon| - \theta\} - \ln\{\mu(\frac{\theta}{1+\theta} - \gamma\,|\epsilon|)\}]$$

$$\leq K\mu^{\mu+\frac{1}{2}}e^{-\mu} \exp[(1+\theta)\mu\gamma\,|\epsilon| - \ln\{\mu^{\frac{1}{2}}(\frac{\theta}{1+\theta} - \gamma\,|\epsilon|)\}]$$

$$\leq \mu^{\mu+\frac{1}{2}}e^{-\mu}O(1/\mu^{\frac{1}{4}}). \tag{2.1.2}$$

For $u \leq 1 - \theta$, change the sign of θ in Eq.(2.1.1). Then

$$\left| \mu^{\mu+1} \int_{0}^{1-\theta} \exp\{\mu(\ln u - u)\} f(\mu\epsilon u) du \right|$$

$$\leq A\mu^{\mu+1} \exp[\mu\ln(1-\theta) - \mu + \mu\{(1-\theta)\gamma\,|\epsilon| + \theta\} - \ln\{\mu(\frac{\theta}{1-\theta} + \gamma\,|\epsilon|)\}]$$

$$\leq \mu^{\mu+\frac{1}{2}}e^{-\mu}O(1/\mu^{\frac{1}{4}}) \tag{2.1.3}$$

again.

From Eq.(2.1.2) and Eq.(2.1.3) can be inferred that the desired integral has been reduced to

$$\mu^{\mu+1} \int_{1-\theta}^{1+\theta} \exp\{\mu(\ln u - u)\} f(\mu\epsilon u) du + \mu^{\mu+\frac{1}{2}}e^{-\mu}\varkappa.$$

In the integral u differs from 1 only by an infinitesimal. Since $\mu\epsilon$ is limited $f(\mu\epsilon u) \simeq f(\mu\epsilon)$ by S-continuity. Therefore, Theorem A.5.2 gives

$$\mu^{\mu+1} \int_{1-\theta}^{1+\theta} \exp\{\mu(\ln u - u)\} f(\mu\epsilon u) du \simeq \mu^{\mu+1} f(\mu\epsilon) \int_{1-\theta}^{1+\theta} \exp\{\mu(\ln u - u)\} du.$$

Substitute $u = 1 + w/\mu^{\frac{1}{2}}$ in the integral. For w limited

$$\mu\{\ln(1 + w/\mu^{\frac{1}{2}}) - w/\mu^{\frac{1}{2}}\} \simeq -w^2/2$$

while $\mu\{\ln(1 + w/\mu^{\frac{1}{2}}) - w/\mu^{\frac{1}{2}}\} < -w^2/4$ for all w in the range of integration. Hence

$$\int_{1-\theta}^{1+\theta} \exp\{\mu(\ln u - u)\} du \simeq \int_{-\mu^{\frac{1}{2}}\theta}^{\mu^{\frac{1}{2}}\theta} \exp(-w^2/2) dw/\mu^{\frac{1}{2}}$$

$$\simeq \int_{-\infty}^{\infty} \exp(-w^2/2) dw/\mu^{\frac{1}{2}}$$

since $\mu^{\frac{1}{2}}\theta$ is unlimited and the integrand is absolutely integrable (cf. Theorem A.5.4). The last integral evaluates to $(2\pi)^{\frac{1}{2}}$ and it has been shown that

$$\int_{0}^{\infty} e^{-t} t^{\mu} f(\epsilon t) dt = \mu^{\mu+\frac{1}{2}} e^{-\mu} f(\mu\epsilon)(1 + \varkappa)(2\pi)^{\frac{1}{2}} \tag{2.1.4}$$

when μ is unlimited.

Now choose $f(t) \equiv 1$ which complies with the conditions of the theorem. Then Eq.(2.1.4) provides *Stirling's formula*

$$\mu! = \mu^{\mu+\frac{1}{2}}e^{-\mu}(2\pi)^{\frac{1}{2}}(1 + \varkappa) \qquad\qquad (2.1.5)$$

for unlimited μ.

Insertion of Stirling's formula into Eq.(2.1.4) recovers the statement of the theorem and the proof is over. ∎

The theorem has been proved for unlimited μ and infinitesimal ϵ subject to $\mu\epsilon$ being limited. In practical applications we would like neither μ to be too large nor ϵ to be too small. Without an estimate for the size of $\varkappa$ in the formula it is difficult to assess what flexibility there is in the choice of μ and ϵ. The matter of estimating errors in asymptotic formulae will be turned to later. For the moment we shall examine how well the formula performs in some examples when the term in $\varkappa$ is omitted. An indication of how valuable asymptotic approximation can be in practice will be obtained thereby.

Inspection of Theorem 2.1.1 discloses that it is not entirely satisfactory despite its generality. When $\mu = 0$ $f(\mu\epsilon)$ becomes $f(0)$ which is independent of ϵ. Theorem 1.7.1 of Chapter 1 indicates that this will be unsatisfactory in general. Therefore, it is desirable to modify Theorem 2.1.1 to allow for some variation of ϵ when $\mu = 0$. If we imagine that, to a first approximation, the variation is due to the first derivative of f we can take care of that by replacing $f(\mu\epsilon)$ with $f((\mu + 1)\epsilon)$. Since f is S-continuous $f((\mu + 1)\epsilon) \simeq f(\mu\epsilon)$ so that the change affects only the error term in the asymptotic approximation. Thus, the following theorem is justified.

Theorem 2.1.1 (Modified) *Under the conditions of Theorem 2.1.1*

$$\int_0^\infty e^{-t}t^\mu f(\epsilon t)dt = \mu!f((\mu + 1)\epsilon)(1 + \varkappa).$$

Example 2.1.1 In Table 2.1.1 is shown the ratio of the two sides of Stirling's formula Eq.(2.1.5) for various values of μ. The departure of the ratio from 1 measures the error in the asymptotic representation of the factorial function. It can be seen that the approximation is pretty good, being only 8% in error at $\mu = 1$, and improves steadily as μ increases. It suggests that, for practical purposes, unlimited numbers can be regarded as going upwards from 10.

Table 2.1.1

μ	Ratio
1	1.08444
5	1.01678
10	1.00837
50	1.00167
100	1.00083

Example 2.1.2 Here the integral

$$I_n = \int_0^\infty \frac{e^{-t}}{(1+\epsilon t)^n} dt$$

is considered. Since $\mu = 0$ Theorem 2.1.1 offers the unhelpful approximation of 1 whatever ϵ or n. Against that, Theorem 1.7.1 of Chapter 1 suggests $1/(1+n\epsilon)$ while Theorem 2.1.1 (Modified) gives $1/(1 + \epsilon)^n$. The two agree for $n = 1$ and are compared with I_1 for various values of ϵ in Table 2.1.2.

Table 2.1.2

ϵ	I_1	$1/(1 + \epsilon)$
0.1	0.91563	0.90909
0.01	0.99019	0.99010
0.005	0.99505	0.99502

Tables 2.1.3 and 2.1.4 compare the exact and two approximate answers for $\epsilon = 0.1$ and $\epsilon = 0.01$ respectively. The approximations perform reasonably well given their simplicity and do better as ϵ decreases. Theorem 1.7.1 is always closer to the correct answer than Theorem 2.1.1 (Modified) which is appreciably better than Theorem 2.1.1. For, although the tables show that I_n approaches 1 as ϵ diminishes it can be some way off even when ϵ is fairly small.

	Table 2.1.3 $\epsilon = 0.1$			Table 2.1.4 $\epsilon = 0.01$		
	Exact	Th. 1.7.1	Th. 2.1.1(M)	Exact	Th. 1.7.1	Th. 2.1.1(M)
I_2	0.84367	0.83333	0.82645	0.98058	0.98039	0.98030
I_3	0.78167	0.76923	0.75131	0.97114	0.97087	0.97059
I_4	0.72778	0.71429	0.68301	0.96188	0.96154	0.96098
I_{10}	0.51218	0.5	0.38554	0.90983	0.90909	0.90529

The tables give the impression that, for practical purposes, infinitesimals should not be treated as exceeding 0.1.

Another feature displayed by Tables 2.1.3 and 2.1.4 is the degradation in the performance of Theorem 2.1.1 (Modified) for I_{10}. It is a demonstration of the necessity to bear in mind the conditions of the theorem. The point is that, with ϵ fixed as n increases, it becomes harder to verify that $(1 + \epsilon)^n$ is S-continuous. Ideally, one might require that $n\epsilon$ does not surpass 0.1 but Table 2.1.3 indicates that this might be unduly restrictive.

We have here an illustration of how asymptotic formulae for the same function which look tolerably similar can produce dissimilar predictions. The source of the difference lies in the error terms which have been neglected. It stresses the importance of keeping an eye on the error terms in asymptotic approximations. **Example 2.1.3** A more general test involving both μ and ϵ is provided by the confluent hypergeometric function. Our purposes will be served by adopting the definition

$$U(\mu+1,\,\mu+p+1,\,1/\epsilon) = \frac{\epsilon^{\mu+1}}{\mu!} \int_0^\infty e^{-t} t^\mu (1+\epsilon t)^{p-1} dt$$

for $\mu > -1$ and $p > 0$. Theorem 2.1.1 gives

$$U \sim \epsilon^{\mu+1}(1+\mu\epsilon)^{p-1}$$

whereas Theorem 2.1.1 (Modified) offers

$$U \sim \epsilon^{\mu+1}\{1+(\mu+1)\epsilon\}^{p-1}.$$

Comparison of the exact U with these approximations is carried out in Tables 2.1.5 and 2.1.6 for $\epsilon = 0.1$ and $\epsilon = 0.01$ respectively over a range of μ and p. Note that in these tables when all the numbers in a row are multiplied by a power of 10 that power is shown at the end of the line in ( ).

Table 2.1.5　　$\epsilon = 0.1$

		Exact	Th. 2.1.1	Th. 2.1.1(M)	
$\mu = 1,$	$p = 1/2$	0.0091718	0.0095346	0.00912871	
	$p = 3/2$	0.010937	0.010488	0.010954	
	$p = 2$	0.012	0.011	0.012	
	$p = 5$	0.02268	0.014641	0.020736	
$\mu = 2,$	$p = 1/2$	0.00088213	0.00091287	0.00087706	
	$p = 3/2$	0.0011378	0.0010954	0.0011402	
	$p = 2$	0.0013	0.0012	0.0013	
	$p = 5$	0.003196	0.0020736	0.0028561	
$\mu = 5,$	$p = 1/2$	7.9717	8.1650	7.9057	(10^{-7})
	$p = 3/2$	1.2614	1.2247	1.2649	(10^{-6})
	$p = 2$	1.6	1.5	1.6	(10^{-6})
	$p = 5$	7.5664	5.0625	6.5536	(10^{-6})
$\mu = 10,$	$p = 1/2$	6.9636	7.0711	6.9007	(10^{-12})
	$p = 3/2$	1.4447	1.4142	1.4491	(10^{-11})
	$p = 2$	2.1	2	2.1	(10^{-11})
	$p = 5$	2.2586	1.6	1.9448	(10^{-10})

The case $p = 2$ is especially interesting because then Theorem 2.1.1 (Modified) reproduces the exact answer while Theorem 2.1.1 is in error by ϵ. This feature is verified by the calculations.

Evidently, both approximations are tolerably accurate for the range of μ and p under consideration. Both improve as ϵ decreases but Theorem 2.1.1 (Modified) is consistently better than Theorem 2.1.1. Some deterioration in performance is observable for $p = 5$ when $\epsilon = 0.1$, exhibiting once again the importance of S-continuity.

Table 2.1.6 $\epsilon = 0.01$

		Exact	Th. 2.1.1	Th. 2.1.1(M)	
$\mu = 1$,	$p = 1/2$	0.000099022	0.000099504	0.000099015	
	$p = 3/2$	0.00010099	0.00010050	0.00010100	
	$p = 2$	0.000102	0.000101	0.000102	
	$p = 5$	0.00010837	0.00010406	0.00010824	
$\mu = 2$,	$p = 1/2$	9.8543	9.9015	9.8533	(10^{-7})
	$p = 3/2$	1.0148	1.00995	1.01489	(10^{-6})
	$p = 2$	1.03	1.02	1.03	(10^{-6})
	$p = 5$	1.1274	1.0824	1.1255	(10^{-6})
$\mu = 5$,	$p = 1/2$	9.7148	9.759	9.7129	(10^{-13})
	$p = 3/2$	1.0295	1.0247	1.0296	(10^{-12})
	$p = 2$	1.06	1.05	1.06	(10^{-12})
	$p = 5$	1.2666	1.2155	1.2625	(10^{-12})
$\mu = 10$,	$p = 1/2$	9.4947	9.5346	9.4916	(10^{-23})
	$p = 3/2$	1.0534	1.0488	1.0536	(10^{-22})
	$p = 2$	1.11	1.1	1.11	(10^{-22})
	$p = 5$	1.5263	1.4641	1.5181	(10^{-22})

Notice that putting $p = \mu$ in this integral to make it a function of μ only could invalidate the conclusions of Theorem 2.1.1. The substitution furnishes an f which is a function of μ and it has been assumed in Theorem 2.1.1 and its modification that f is independent of μ.

One further comment on Theorem 2.1.1 is pertinent. The proof when μ is unlimited showed that the main contribution came from u in a neighbourhood of 1. This fact can be used to estimate integrals over ranges other than from 0 to ∞.

Theorem 2.1.2 *If μ is unlimited but $\mu\epsilon$ is limited*

$$\int_a^b e^{-t}t^\mu f(\epsilon t)dt = \mu! f(\mu\epsilon)(1 + \varkappa)$$

when $0 \leq a \leq \mu - \mu^{\frac{3}{4}}$ and $b \geq \mu + \mu^{\frac{3}{4}}$.

The modified version is available also if preferred.

2.2   Laplace integrals

Integrals of the type

$$F(x) = \int_0^\infty e^{-xt} f(t)dt,$$

where x is positive, are known as *Laplace integrals*. Their behaviour as $x \to \infty$ can be deduced from the preceding section by the change of variable $t = u/x$ whence

$$F(x) = \int_0^\infty e^{-u} f(u/x)du/x.$$

Theorem 2.2.1 *If x is unlimited*

$$\begin{aligned} F(x) &= f(0)(1 + \text{и})/x \\ &= f(1/x)(1 + \text{и})/x \ \ \textit{(Modified)} \end{aligned}$$

provided that $f(t)$ is S-continuous at $t = 0$ and $|f(t)| < Ae^{\gamma t}$ for $t \geq 0$ with A and γ standard constants.

If $d > 0$ is standard

$$\left| \int_d^\infty e^{-xt} f(t)dt \right| < A \int_d^\infty e^{-(x-\gamma)t} dt < KAe^{-(x-\gamma)d} \tag{2.2.1}$$

for unlimited x. The inequality in Eq.(2.2.1) demonstrates that the integral over (d, ∞) is exponentially small and could have been dropped without affecting Theorem 2.2.1. In other words Theorem 2.2.1 also supplies an estimate for $\int_0^d$.

From now on the modified version of the asymptotic formula will be omitted and the reader left to supply the missing detail.

Theorem 2.2.1 was based on earlier theory but there is an alternative proof which is capable of useful generalisation.

Theorem 2.2.2 *Let z be unlimited with $|\text{ph } z| \leq \frac{1}{2}\pi - \delta$ where $\delta > 0$ is standard. If cz is unlimited or $c = \infty$, then*

$$\int_0^c e^{-zt} f(t)dt = f(0)(1 + \text{и})/z$$

when $f(t)$ is S-continuous at $t = 0$ and $|f(t)| < Ae^{\gamma t}$ for $t \geq 0$ with A and γ standard constants.
Proof. For $0 < d < c$

$$\left| \int_d^c e^{-zt} f(t)dt \right| < KA[\exp\{-(\mathcal{R}z - \gamma)d\} - \exp\{-(\mathcal{R}z - \gamma)c\}] \tag{2.2.2}$$

as in Eq.(2.2.1) since $\mathcal{R}z \geq |z| \sin \delta$ which is unlimited. Choose d some infinitesimal but such that zd is unlimited. The right-hand side of Eq.(2.2.2) is then exponentially small and smaller than any inverse power of z. Also

$$\int_0^d e^{-zt} f(t)dt \simeq f(0) \int_0^d e^{-zt} dt \simeq f(0) \int_0^\infty e^{-zt} dt \tag{2.2.3}$$

by a repetition of the argument leading to Eq.(2.2.2). The theorem follows at once. ∎

When it is known that $f(t)$ is regular in some angular sector including the positive real axis Theorem 2.2.2 can be extended by analytic continuation when $c = \infty$. Suppose, in fact, that $f(t)$ is regular in $\alpha_1 \leq \text{ph}\, t \leq \alpha_2$ with $\alpha_1 \leq 0$ and $\alpha_2 \geq 0$, one at least being non-zero. Suppose too that $|f(t)| < Ae^{\gamma|t|}$ for the same range of ph t. The function $F(z)$ is regular in z for $|\text{ph}\, z| \leq \frac{1}{2}\pi - \delta$. The aim is to continue it analytically by moving the contour of integration. Let $-\frac{1}{2}\pi + \delta \leq \text{ph}\, z \leq -\delta$ (it will be assumed that δ is small enough for this and subsequent operations). Pick β to be the smaller of $\frac{1}{2}\pi$ and α_2. By Cauchy's theorem

$$\int_0^\infty e^{-zt}f(t)dt = \int_0^{Re^{i\beta}} e^{-zt}f(t)dt - \int_C e^{-zt}f(t)dt$$

(see Fig. 2.2.1). On the circular arc C, $|\text{ph}(zt)| \leq \frac{1}{2}\pi - \delta$ and

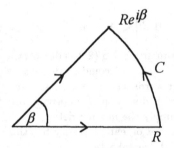

Figure 2.2.1 Moving the contour

$$\left| \int_C e^{-zt}f(t)dt \right| \leq AR\beta \exp(-|z|\,R\sin\delta + \gamma R)$$

which tends to zero as $R \to \infty$ so long as $|z|\sin\delta > \gamma$. Hence

$$F(z) = \int_0^{\infty e^{i\beta}} e^{-zt}f(t)dt. \tag{2.2.4}$$

The integral on the right of Eq.(2.2.4) is regular in z for $-\frac{1}{2}\pi - \beta + \delta \leq \text{ph}\, z \leq \frac{1}{2}\pi - \beta - \delta$ ($|z|\sin\delta > \gamma$). Since β does not exceed $\frac{1}{2}\pi$ this region has an overlap with $|\text{ph}(z)| \leq \frac{1}{2}\pi - \delta$ and so the integral provides an analytic continuation of $F(z)$ into the new region. If $\beta \neq \alpha_2$ the exercise can be repeated starting from the new position of the contour and moving it by $\pi/2$ or up to α_2. Obviously, the process can be carried on until α_2 is reached. Similarly, the contour can rotate in the opposite direction until α_1 is attained.

Theorem 2.2.3 Let $f(t)$ be regular in $\alpha_1 \leq \mathrm{ph}\, t \leq \alpha_2$ ($\alpha_1 \leq 0$, $\alpha_2 \geq 0$) with $|f(t)| < Ae^{\gamma|t|}$ in this sector and S-continuous at $t = 0$. Then $F(z)$ and its analytic continuation satisfy

$$F(z) = f(0)(1 + \varkappa)/z$$

for unlimited z with $-\frac{1}{2}\pi - \alpha_2 + \delta \leq \mathrm{ph}\, z \leq \frac{1}{2}\pi - \alpha_1 - \delta.$

Remark that, if $\alpha_2 - \alpha_1 > \pi$, $\mathrm{ph}\, z$ can range over an angle greater than 2π. That does not mean that, when z is altered to $ze^{2\pi i}$, $F(z)$ is the same because the analytic continuation may not be single-valued. Nevertheless, the asymptotic behaviour of the analytic continuation is unchanged.

It may happen that $f(t)$ vanishes at $t = 0$, in which case the preceding theorems are not of much use. However, an examination of the proof of Theorem 2.2.2 reveals that the specific form of f is employed only in an infinitesimal neighbourhood of the origin. This observation enables a more general behaviour to be encompassed.

Theorem 2.2.4 Let z be unlimited with $|\mathrm{ph}\, z| \leq \frac{1}{2}\pi - \delta$ where $\delta > 0$ is standard. Let $f(t) \simeq f_0(t)$ when $t \simeq 0$. If cz is unlimited or $c = \infty$, then

$$\int_0^c e^{-zt} f(t) dt = (1 + \varkappa) \int_0^\infty e^{-zt} f_0(t) dt$$

when $|f(t)|$ and $|f_0(t)|$ are bounded by $Ae^{\gamma t}$ for $t > 0$ with A and γ standard.
Proof. The sole change to the proof of Theorem 2.2.2 is in Eq.(2.2.3) which becomes

$$\int_0^d e^{-zt} f(t) dt \simeq \int_0^d e^{-zt} f_0(t) dt \simeq \int_0^\infty e^{-zt} f_0(t) dt. \qquad \blacksquare$$

A typical example occurs when $f_0(t) = t^\lambda$ with $\mathcal{R}\lambda \geq 0$. Then

$$\int_0^c e^{-zt} f(t) dt = \lambda! (1 + \varkappa)/z^{\lambda+1}$$

where the principal value of the power is taken when necessary.

A slightly more general version of Theorem 2.2.4 deals with the case when the information about f is given on a ray in the complex plane rather than the positive real axis.

Corollary 2.2.4 When $|f(te^{i\alpha})|$ and $|f_0(te^{i\alpha})|$ are bounded by $Ae^{\gamma t}$ for $t \geq 0$ and $f(t) \simeq f_0(t)$ for $t \simeq 0$ on the ray

$$\int_0^{\infty e^{i\alpha}} e^{-zt} f(t) dt = (1 + \varkappa) \int_0^{\infty e^{i\alpha}} e^{-zt} f_0(t) dt$$

for z unlimited with $|\alpha + \mathrm{ph}\, z| \leq \frac{1}{2}\pi - \delta.$

Proof. In the integral on the left put $t = ue^{i\alpha}$. Then Theorem 2.2.4 is applicable. The substitution $u = te^{-i\alpha}$ recovers the integral on the right. ∎

At first sight the exclusion of an integrable singular f by the conditions of Theorem 2.2.4 appears to be an undesirable limitation. In practice, it occasions no difficulty usually since the singularity can be extracted often in a straight-forward manner. Suppose, in fact, that $f(t) - f_1(t)$ satisfies the same conditions as f in Theorem 2.2.4. Apply Theorem 2.2.4 to $f(t) - f_1(t)$ with the result

$$\int_0^c e^{-zt} f(t) dt = \int_0^c e^{-zt} f_1(t) dt + (1 + \varkappa) \int_0^\infty e^{-zt} f_0(t) dt. \qquad (2.2.5)$$

The upper limit on the integral of f_1 can be replaced by ∞ with infinitesimal error if f_1 is bounded by an exponential for $t \geq c$. The above device allows for f having singular behaviour like $\ln t$ or t^λ $(0 > \lambda > -1)$ at the origin.

2.3   Laplace-type integrals

Let $h(t)$ be a real-valued function for real t. When a and b are real

$$\int_a^b f(t) e^{-zh(t)} dt$$

is called a *Laplace-type integral.* If h is strictly monotonic on the interval of integration, say h increases steadily with t, substitution of $h(t) - h(a) = u$ converts the integral into a Laplace integral which can be handled by the methods of the preceding section.

A new situation arises if the derivative of h vanishes at some point of the interval of integration or becomes infinite there. When there is a finite number of such points the interval may be segmented into a finite number of sub-intervals. In each sub-interval there is one such point at one end and no other. A simple change of variable puts the end of the sub-interval at the origin. Hence, it is sufficient to consider integrals of the type

$$I(z) = \int_0^c f(t) e^{-zh(t)} dt$$

where $h'(t)$ does not change sign throughout the integration. To fix ideas it will be assumed that $h'(t) > 0$ for $0 < t \leq c$.

Introduce the new variable $u = h(t) - h(0)$. Then

$$I(z) = e^{-zh(0)} \int_0^C \frac{f(t)}{h'(t)} e^{-zu} du$$

where $C = h(c) - h(0)$. So long as Cz is unlimited and $f(t)/h'(t)$ is bounded by an exponential in u (which will be assumed henceforth) the theory of the preceding section is applicable. According to that the main contribution to $I(z)$ comes from the integrand near $u = 0$.

It is not possible to proceed much further without making some assumption about the relation between t and u. For definiteness it will be supposed that $h'(t) \simeq a_0 t^{\mu-1}$ with $a_0 > 0$ and $\mu > 0$ when $t \simeq 0$. Then $u \simeq a_0 t^\mu / \mu$ whence $t \simeq (\mu u / a_0)^{1/\mu}$. Accordingly

$$\frac{f(t)}{h'(t)} \simeq \frac{f\{(\mu u/a_0)^{1/\mu}\}}{a_0(\mu u/a_0)^{1-1/\mu}}.$$

What happens now depends upon the behaviour of f. If $f(t) \simeq b_0 t^{\lambda-1}$ for $t \simeq 0$ and $\lambda \geq \mu$

$$I(z) = \left(\frac{\lambda}{\mu} - 1\right)! \frac{b_0}{\mu} e^{-zh(0)} \left(\frac{\mu}{a_0 z}\right)^{\lambda/\mu} (1 + \varkappa) \qquad (2.3.1)$$

for $|\mathrm{ph}\, z| \leq \frac{1}{2}\pi - \delta$, $\delta > 0$. When $0 < \lambda < \mu$ recourse to a device like Eq.(2.2.5) can supply a similar result.

The formula of Eq.(2.3.1) permits many different kinds of conduct by f and h near the origin. Of particular interest is the case when h is stationary at the endpoint. Then $\mu = 2$ and $a_0 = h''(0)$. If $f(t) \simeq b_0 + b_1 t$ for $t \simeq 0$, deployment of Eq.(2.2.5) gives

$$\int_0^c f(t) e^{-zh(t)} dt = f(0) e^{-zh(0)} \left\{ \frac{\pi}{2zh''(0)} \right\}^{\frac{1}{2}} (1 + \varkappa). \qquad (2.3.2)$$

2.4   Complex contours

An integral more general than the Laplace-type is

$$\int_a^b f(w) e^{-zh(w)} dw$$

where a, b are complex numbers and the path of integration is some curve in the complex plane. Discussion of this integral is much more complicated than for the Laplace-type and can involve a good deal of manipulation. The object of the manipulation is to convert as far as possible the integral into one of Laplace-type so that the theory of Section 2.3 can be invoked. This means looking for points where h is stationary or is not regular and trying to deform the contour so that it passes through some or all of these points.

Points where $h'(w) = 0$ are known as *saddle points*. The curves of interest through saddle points will be those on which $\mathcal{I}h(w)$ is constant for then Section

2.3 can be applied directly. Not all such curves are suitable because on some of them $\mathcal{R}h$ increases on leaving the saddle point whereas on others it decreases. The ones that are needed are those on which $\mathcal{R}h$ increases or $-\mathcal{R}h$ decreases. For this reason the required curves are called *curves of steepest descent*. If the original path of integration can be deformed into a curve or curves of steepest descent the evaluation of the integral is said to be tackled by the *method of steepest descents*. Determination of the curves of steepest descent throughout the complex plane can be a formidable matter. Yet a good idea of their position is essential in order to settle whether deformation of the original contour is feasible and whether any singularities of $f(w)$ have to be circumnavigated in the deformation. There can be no assumption that an automatic transference to curves of steepest descent will be possible or satisfactory. Other paths of integration may be more convenient—it all depends on the forms of h and f.

The foregoing remarks indicate that it will be extremely difficult to formulate conditions which cover a generality of circumstances. Therefore, we shall assume that some preliminary manipulation has been undertaken so that the integral is of the type about to be described.

It will be assumed that a and b lie in an open domain D in which f and h are single-valued and regular. Further, a will be taken to be limited whereas b can be finite or infinite. The path of integration is supposed to lie in D. With regard to z observe that $zh(w) = z\exp(-i\,\mathrm{ph}\,z)\{h(w)\exp(i\,\mathrm{ph}\,z)\}$. So, by adjustment of h if necessary, it is always possible to replace z by the positive x and from now on that adjustment will be assumed to have been made. The behaviour of f and h for $w \simeq a$ is specified by

$$
\begin{aligned}
f(w) &\simeq b_0(w-a)^{\lambda-1}, \\
h(w) &\simeq h(a) + a_0(w-a)^{\mu}, \\
h'(w) &\simeq \mu a_0(w-a)^{\mu-1}
\end{aligned}
$$

with $\mu > 0$, $\mathcal{R}(\lambda) > 0$ and $a_0 \neq 0$ but not infinitesimal. If either λ or μ is not an integer a is forced to be a boundary point of D to comply with the requisite regularity of f and h.

It will be assumed that the path of integration is such that

$$\mathcal{R}\{h(w) - h(a)\} > 0 \qquad\qquad (2.4.1)$$

except possibly at an endpoint. Also it will be supposed that, as $w \to b$ along the path, $\mathcal{R}\{h(w) - h(a)\}$ is bounded away from zero. The integral will be taken to be absolutely convergent.

Because f and h can have branch points at $w = a$ considerable care has to be devoted to defining precisely the branches of the multiple-valued functions which arise. Let θ_0 be the angle of the slope of the path of integration at $w = a$

i.e.

$$\theta_0 = \lim_{w \to a} \mathrm{ph}(w - a),$$

the limiting process being carried out with w on the path of integration. Once θ_0 has been settled the branches of f and h are defined by

$$(w - a)^{\lambda-1} = |w - a|^{\lambda-1} e^{i(\lambda-1)\theta_0}, \ (w - a)^{\mu} = |w - a|^{\mu} e^{i\mu\theta_0}$$

for w near a and by continuity elsewhere. The phase of a_0 is involved also. Both $\mathrm{ph}\,a_0$ and θ_0 can be altered by multiples of 2π but, to meet Eq.(2.4.1), there must be a choice such that

$$|\mu\theta_0 + \mathrm{ph}\,a_0| \leq \pi/2. \tag{2.4.2}$$

Once values of θ_0 and $\mathrm{ph}\,a_0$ have been found to satisfy this criterion they are invariable from this point onward.

The plan now is to split the contour at a point c, yet to be fixed, convert the integral from a to c into Laplace-type and estimate the error due to the integral from c to b.

Make the change of variable

$$h(w) - h(a) = u$$

with the understanding that $\mathrm{ph}\,u \to \mu\theta_0 + \mathrm{ph}\,a_0$ as $w \to a$ along the path of integration. Elsewhere the phase is determined by continuity and is unique on the path of integration because Eq.(2.4.1) does not permit u to be zero there. On account of Eq.(2.4.1) and Eq.(2.4.2), when $u \neq 0$,

$$|\mathrm{ph}\,u| < \pi/2 \tag{2.4.3}$$

which prevents u from straying outside a single Riemann sheet as w travels along the contour of integration.

In the u-plane the path of integration is a contour starting from the origin. In general, the contour will not be the desired straight line. To organise this, near the origin at any rate, we move to another plane where our objective can be achieved. Put $v^{\mu} = u$ with $\mathrm{ph}\,v = (\mathrm{ph}\,u)/\mu$. As w approaches a, $v \simeq (w-a)a_0^{1/\mu}$. Thus, to a first approximation, small circles centred on $w = a$ in the w-plane map into circles, with centre the origin, in the v-plane. They will be perturbed somewhat when the full formula connecting v and w is employed. We infer that there is a $\delta > 0$ such that the domain $|w - a| < \delta$ is mapped conformally onto a domain V of the v-plane. There is now a $c \neq a$ for which the domain $|v| \leq |h(c) - h(a)|^{1/\mu}$ is contained in V. Within this circle the path of integration can be shifted into the straight line joining the origin and $\{h(c) - h(a)\}^{1/\mu}$. A straight line in the v-plane maps into a straight line in the u-plane. Hence

$$\int_a^c f(w)e^{-xh(w)}dw = e^{-xh(a)} \int_0^C \frac{f(w)}{h'(w)} e^{-xu} du$$

with $C = h(c) - h(a)$. Substitute $u = t \exp(i \,\mathrm{ph}\, C)$ where, in accordance with Eq.(2.4.3), $|\mathrm{ph}\, C| < \pi/2$. Then

$$\int_a^c f(w) e^{-xh(w)} dw = e^{-xh(a)+i\,\mathrm{ph}\, C} \int_0^{|C|} \frac{f(w)}{h'(w)} e^{-Zt} dt$$

where $Z = x e^{i\,\mathrm{ph}\, C}$ and $\mathrm{ph}\, Z = \mathrm{ph}\, C$. The integral on the right is of Laplace-type and can be handled as in Section 2.3.

Quoting from Eq.(2.3.1) we have, after including the various changes of variable

$$\int_a^c f(w) e^{-xh(w)} dw = \left(\frac{\lambda}{\mu} - 1\right)! \frac{b_0 e^{-xh(a)}}{\mu(a_0 x)^{\lambda/\mu}} (1 + \varkappa). \qquad (2.4.4)$$

There remains the matter of assessing the contribution of the other piece of the integral. From Eq.(2.4.1) and the condition on $h(w) - h(a)$ there is a positive ζ such that

$$\mathcal{R}\{h(w) - h(a)\} \geq \zeta > 0$$

on the path of integration. Choose some limited positive number x_0. Then

$$\mathcal{R}x\{h(w) - h(a)\} \geq (x - x_0)\zeta + \mathcal{R}x_0\{h(w) - h(a)\}.$$

Hence

$$\left| \int_c^b e^{-x\{h(w)-h(a)\}} f(w) dw \right| \leq \exp\{-(x - x_0)\zeta + \mathcal{R}x_0 h(a)\}$$
$$\times \int_c^b \exp\{-\mathcal{R}x_0 h(w)\} f(w) dw.$$

The integral is bounded by virtue of the assumption on convergence. The other factors on the right-hand side are constant or decay exponentially as x increases. Therefore, the correction to Eq.(2.4.4) is infinitesimal and

$$\int_a^b f(w) e^{-xh(w)} dw = \left(\frac{\lambda}{\mu} - 1\right)! \frac{b_0 e^{-xh(a)}}{\mu(a_0 x)^{\lambda/\mu}} (1 + \varkappa). \qquad (2.4.5)$$

The essential feature of this method which enables the quoting of Eq.(2.4.5) is that $\mathcal{R}h(w)$ attains its smallest value on the path of integration at $w = a$. Consequently, in applications, the aim will be to shift contours to such paths. Often it will not be easy to find suitable paths though locating any saddle points should be of great assistance generally. Actually, it is sufficient to confirm the existence of a suitable path, possibly by an investigation of the mapping between the w- and v-planes, without knowing its precise position unless singularities of f intervene.

Example 2.4.1 Consider

$$\int_a^b e^{-xh(w)} dw$$

with a, b real $(b > a)$ and

$$h(w) = a_0 i(w - a) + a_1 (w - a)^2$$

where $a_0 > 0$, $a_1 > 0$. The function h satisfies the conditions imposed above. Also $\lambda = 1$, $\mu = 1$ and $\text{ph}(a_0 i) = \pi/2$. Moreover, Eq.(2.4.2) holds with $\theta_0 = 0$. Hence, from Eq.(2.4.5),

$$\int_a^b e^{-xh(w)} dw = \frac{1 + \varkappa}{a_0 ix}. \qquad (2.4.6)$$

For the same integrand taken from d to a with $d < a$ write

$$\int_d^a e^{-xh(w)} dw = -\int_a^d e^{-xh(w)} dw.$$

Now θ_0 has to be changed by π. If $\theta_0 = -\pi$ is chosen consistency with Eq.(2.4.2) is achieved without any alteration to the phase of $a_0 i$. Thus

$$\int_d^a e^{-xh(w)} dw = -\frac{1 + \varkappa}{a_0 ix}. \qquad (2.4.7)$$

The combination of Eq.(2.4.6) and Eq.(2.4.7) leads to

$$\int_d^b e^{-xh(w)} dw = \varkappa / a_0 ix$$

which is not very illuminating usually.

Example 2.4.2 Here the same integral as in Example 2.4.1 is studied but with $a_0 = 0$. In this case $\mu = 2$ and $\theta_0 = 0$, $\text{ph}\, a_1 = 0$ comply with Eq.(2.4.2). Hence

$$\int_a^b e^{-xh(w)} dw = \frac{(1 + \varkappa)\pi^{\frac{1}{2}}}{2(a_1 x)^{\frac{1}{2}}}. \qquad (2.4.8)$$

For the integral from d to a, $\theta_0 = -\pi$ forces $\text{ph}\, a_1 = 2\pi$ to meet Eq.(2.4.2). Therefore

$$\int_d^a e^{-xh(w)} dw = -\frac{(1 + \varkappa)\pi^{\frac{1}{2}}}{2(a_1 x)^{\frac{1}{2}} e^{\pi i}}. \qquad (2.4.9)$$

Combining Eq.(2.4.8) and Eq.(2.4.9) we have

$$\int_d^b e^{-xh(w)} dw = (1 + \varkappa)\pi^{\frac{1}{2}} / (a_1 x)^{\frac{1}{2}}. \qquad (2.4.10)$$

These two examples demonstrate the importance of checking that Eq.(2.4.2) is verified before any substitutions are made in Eq.(2.4.5). Otherwise it is easy to arrive at incorrect conclusions about the performance of an integral.

In the next example the initial position of the contour does not permit the immediate application of the foregoing theory and it has to be deformed into one that meets the constraints which have been imposed.

Example 2.4.3 Let

$$I(x) = \int_{-\infty}^{\infty} e^{-xh(w)} dw$$

where $h(w) = w^2 + 2\ln w$. The principal branch of the logarithm is selected with the branch line along the negative real axis. The path of integration is to pass above the branch line so that $\ln w = \ln|w| + \pi i$ for $w < 0$.

The function $h(w)$ does not satisfy our conditions on the real axis; it swings between $\pm\infty$ as w varies without having a minimum. So look for points where $h'(w) = 0$; these are found easily to be $w = \pm i$. Try first $w = i$ so as to avoid having to loop round the branch line. Put $w = i + t$ with t real to see if a line parallel to the real axis is appropriate. Then

$$h(w) = t^2 - 1 + 2it + 2\ln(i + t)$$

and

$$\mathcal{R}h(w) = t^2 - 1 + \ln(t^2 + 1).$$

As t increases from $-\infty$, $\mathcal{R}h(w)$ decreases steadily until it reaches -1 at $t = 0$ and then increases again. Consequently, a suitable contour has been discovered. There is no problem in deforming the original contour to this new position by Cauchy's theorem because the w^2 ensures that any pieces at infinity give no contribution. Near $t = 0$, expansion of the logarithm gives

$$h(w) \simeq -1 + 2\ln i + 2t^2 = \pi i - 1 + 2t^2$$

on account of the choice of branch for the logarithm. Invoking Eq.(2.4.10) with $a_1 = 2$ we have

$$I(x) = \left(\frac{\pi}{2x}\right)^{\frac{1}{2}} (1 + \text{и}) e^{x(1-\pi i)}.$$

2.5   Fourier integrals

Most of the preceding theory is irrelevant to Fourier integrals of the type

$$\int_a^b e^{-ixt} f(t) dt$$

with x real because the exponential decay with increasing x is absent. Consequently, general results are harder to come by and more effort has gone into dealing with particular forms for f. One theorem which does not refer to the specific behaviour of f will now be stated.

Theorem 2.5.1 (Riemann-Lebesgue Lemma) *If $\int_a^b |f(t)| \, dt$ exists then*

$$\int_a^b e^{-ixt} f(t) dt \simeq 0$$

for unlimited x ($b = \infty$ is permitted and so is $a = -\infty$).

Proof. Since $f(t)$ could always be defined to be zero outside (a, b) there is no loss of generality in taking $a = -\infty$ and $b = \infty$. Now

$$
\begin{aligned}
\int_{-\infty}^{\infty} e^{-ixt} f(t)dt &= -\int_{-\infty}^{\infty} e^{-ixt+\pi i} f(t)dt \\
&= -\int_{-\infty}^{\infty} e^{-ixt} f(t - \pi/x)dt \\
&= \frac{1}{2}\int_{-\infty}^{\infty} e^{-ixt}\{f(t) - f(t - \pi/x)\}dt.
\end{aligned}
$$

Also

$$
\left|\int_{-\infty}^{\infty} e^{-ixt}\{f(t) - f(t - \pi/x)\}dt\right| \le \int_{-\infty}^{\infty} |f(t) - f(t - \pi/x)|\, dt \simeq 0
$$

by Corollary A.5.4 since π/x is infinitesimal. The theorem is proved. ∎

Rather more detail about the infinitesimal can be obtained by assuming something about the structure of f. Suppose, in fact, that $f = \psi_1 - \psi_2$ where ψ_1, ψ_2 are integrable functions which are non-increasing and positive. They are to be bounded except possibly at $t = a$ but, as $t \to a + 0$, their growth is restricted by

$$
\psi_1(t),\ \psi_2(t) = \text{н}\,(t - a)^{-1/p} \tag{2.5.1}
$$

with $1 < p < \infty$.

Corollary 2.5.1 *Let a be limited and $\int_a^b |f(t)|\, dt$ exist with f having the structure just described. Then*

$$
\int_a^b e^{-ixt} f(t)dt = \text{н}\,|x|^{-1+1/p}
$$

for unlimited x.

Proof. By the second mean value theorem for integrals

$$
\begin{aligned}
\int_{a+1/|x|}^b \psi_1(t) e^{-ixt} dt &= \psi_1(a + 1/|x| + 0)\int_{a+1/|x|}^{\xi} e^{-ixt} dt \\
&= \psi_1(a + 1/|x| + 0)(e^{-iax - i\,\mathrm{sgn}\,x} - e^{-ix\xi})/ix
\end{aligned}
$$

where $a + 1/|x| < \xi < b$ with b finite. From Eq.(2.5.1) the right-hand side can be expressed as $\text{н}\,|x|^{-1+1/p}$ and the same is true for the integral of ψ_2. Furthermore

$$
\left|\int_a^{a+1/|x|} \psi_1(t) e^{-ixt} dt\right| < \int_a^{a+1/|x|} \psi_1(t)dt = \text{н}\,|x|^{-1+1/p}
$$

from Eq.(2.5.1). Thus, the corollary has been verified for finite b.

If b is infinite the same estimate is relevant by replacing b above by ω (ω unlimited) so long as ω can be chosen so that $\int_\omega^\infty |f(t)|\, dt$ offers the same contribution. ∎

Corollary 2.5.1a *Under the conditions of Corollary 2.5.1 but with f of bounded variation*

$$\int_a^b e^{-ixt} f(t)dt = K \, |x|^{-1}$$

where K is limited.

Proof. The argument follows the same lines as Corollary 2.5.1 but now ψ_1 and ψ_2 are bounded at $t = a$. ∎

Another useful theorem is

Theorem 2.5.2 *If $f, f', \ldots, f^{(N)}$ are all absolutely integrable*

$$\int_a^b e^{-ixt} f(t)dt = \sum_{p=0}^{N-1} \{f^{(p)}(a)e^{-iax} - f^{(p)}(b)e^{-ibx}\}/(ix)^{p+1} + \varkappa/x^N$$

where $f^{(p)}(b)$ is replaced by zero if $b = \infty$.

It is understood here and subsequently that when a power of ix is involved the meaning to be attached to ix is

$$ix = |x| \exp\left(i\frac{\pi}{2}\operatorname{sgn} x\right).$$

Likewise $-ix$ means $|x| \exp\{-i(\pi/2)\operatorname{sgn} x\}$.

Proof. Integration by parts gives

$$\int_a^b e^{-ixt} f(t)dt = \{f(a)e^{-iax} - f(b)e^{-ibx}\}/ix + \int_a^b e^{-ixt} f'(t)dt/ix. \qquad (2.5.2)$$

Repeated application leads to an integral of $f^{(N)}(t)$. The Riemann-Lebesgue Lemma (Theorem 2.5.1) then shows that the integral is infinitesimal. The stated expansion is confirmed.

If the integrals in Eq.(2.5.2) continue to exist as $b \to \infty$ it is necessary that $f(b) \to 0$. The same argument deals with $f^{(p)}(b)$ and the proof is terminated. ∎

Notice that, if f and its $(N-1)$ derivatives vanish at both a and b, it can be deduced immediately that the Fourier integral is $\varkappa/x^N$.

Very often a convenient aid to deriving formulae for Fourier integrals is

$$\int_a^b e^{-ixt} f(t)dt = \lim_{\epsilon \to +0} \int_a^b e^{-(\epsilon+ix)t} f(t)dt \qquad (2.5.3)$$

when $b > a$, a is limited and f absolutely integrable. The advantage of Eq.(2.5.3) in proving results is that the limit can exist even when f is not absolutely integrable.

For example, when $\mu > 0$,

$$\int_a^\infty (t-a)^{\mu-1} e^{-(\epsilon+ix)t} dt = (\mu-1)! e^{-(\epsilon+ix)a}/(\epsilon+ix)^\mu.$$

Therefore

$$\lim_{\epsilon \to +0} \int_a^\infty (t-a)^{\mu-1} e^{-(\epsilon+ix)t} dt = (\mu-1)! e^{-ixa}/(ix)^\mu. \qquad (2.5.4)$$

Another example is

$$\int_0^\infty (t+c)^{\mu-1} e^{-(\epsilon+ix)t} dt$$

where $c > 0$. Repetition of integration by parts supplies

$$\int_0^\infty (t+c)^{\mu-1} e^{-(\epsilon+ix)t} dt = \sum_{p=0}^{M-1} \frac{1}{(\epsilon+ix)^{p+1}} \frac{d^p}{dc^p} c^{\mu-1}$$
$$+ \frac{1}{(\epsilon+ix)^M} \int_0^\infty \frac{d^M}{dc^M} (t+c)^{\mu-1} e^{-(\epsilon+ix)t} dt.$$

The integral converges absolutely even if $\epsilon = 0$ when $M > \mu$. Hence

$$\lim_{\epsilon \to +0} \int_0^\infty (t+c)^{\mu-1} e^{-(\epsilon+ix)t} dt = \sum_{p=0}^{M-1} \frac{1}{(ix)^{p+1}} \frac{d^p}{dc^p} c^{\mu-1} + \frac{и}{x^M} \qquad (2.5.5)$$

for any $M > \mu$. If μ is the positive integer m the process terminates and the infinitesimal is absent i.e.

$$\lim_{\epsilon \to +0} \int_0^\infty (t+c)^{m-1} e^{-(\epsilon+ix)t} dt = \sum_{p=0}^{m-1} \frac{1}{(ix)^{p+1}} \frac{d^p}{dc^p} c^{m-1}. \qquad (2.5.6)$$

A different way of writing Eq.(2.5.5) comes from drawing benefit from Eq.(2.5.4). Thus

$$\lim_{\epsilon \to +0} \int_0^\infty (t+c)^{\mu-1} e^{-(\epsilon+ix)t} dt = \lim_{\epsilon \to +0} \int_0^\infty \sum_{p=0}^{M-1} \frac{t^p}{p!} \frac{d^p}{dc^p} c^{\mu-1} e^{-(\epsilon+ix)t} dt + \frac{и}{x^M} \qquad (2.5.7)$$

which shows that the terms in the series of Eq.(2.5.5) originate from the (formal) expansion of $(t+c)^{\mu-1}$ in powers of t, even though that expansion is strictly valid only for $|t| < c$.

Example 2.5.1 The integral to be discussed is

$$I(x) = \int_a^b (t-a)^{\mu-1} e^{-ixt} dt,$$

a and b being limited with $b > a$ while $\mu > 0$. In this case it is certainly true that

$$I(x) = \lim_{\epsilon \to +0} \int_a^b (t-a)^{\mu-1} e^{-(\epsilon+ix)t} dt.$$

Now

$$\int_a^b (t-a)^{\mu-1} e^{-(\epsilon+ix)t} dt = \int_a^\infty (t-a)^{\mu-1} e^{-(\epsilon+ix)t} dt - \int_b^\infty (t-a)^{\mu-1} e^{-(\epsilon+ix)t} dt.$$

The first integral on the right has a limit by Eq.(2.5.4). In the second integral put $t = b + u$ to obtain an integral for which Eq.(2.5.5) can be invoked with $c = b - a$. Hence

$$I(x) = (\mu - 1)!\frac{e^{-iax}}{(ix)^{\mu}} - e^{-ibx}\sum_{p=0}^{M-1}\frac{1}{(ix)^{p+1}}\frac{d^p}{db^p}(b-a)^{\mu-1} + \frac{\unicode{x043A}}{x^M} \qquad (2.5.8)$$

for $M > \mu$, Eq.(2.5.6) being relevant if $\mu = m$.

Note that we could infer also that

$$\lim_{\epsilon \to +0}\int_a^b (t-a)^{\mu-1}e^{-(\epsilon+ix)t}dt + \int_b^{\infty}\sum_{p=0}^{M-1}\frac{(t-b)^p}{p!}\frac{d^p}{db^p}(b-a)^{\mu-1}e^{-(\epsilon+ix)t}dt$$

$$= (\mu-1)!\frac{e^{-iax}}{(ix)^{\mu}} + \frac{\unicode{x043A}}{x^M} \qquad (2.5.9)$$

as for Eq.(2.5.6). Actually this is a kind of generalisation of Theorem 2.5.2 because the left-hand side is the integral from a to ∞ of a function with M integrable derivatives when $\epsilon \neq 0$ if the singularity at $t = a$ is ignored.

Example 2.5.2 Here we study

$$I(x) = \int_a^b (t-a)^{\mu-1}f(t)e^{-ixt}dt$$

with $0 < \mu \leq 1$. Again $I(x)$ can be calculated as a limit after inserting the exponential factor under the conditions that are going to be imposed.

Suppose firstly that f possesses an integrable derivative. Write the integral with the exponential factor as

$$\int_a^b (t-a)^{\mu-1}\{f(t)-f(a)\}e^{-(\epsilon+ix)t}dt + \int_b^{\infty}(b-a)^{\mu-1}\{f(b)-f(a)\}e^{-(\epsilon+ix)t}dt$$

$$+f(a)\int_a^b (t-a)^{\mu-1}e^{-(\epsilon+ix)t}dt - (b-a)^{\mu-1}\{f(b)-f(a)\}\int_b^{\infty}e^{-(\epsilon+ix)t}dt.$$

The first two integrals combine to give an integrand which is differentiable from 0 to ∞ and so, by Theorem 2.5.2, make a contribution of $\unicode{x043A}/x$. The two remaining integrals with $f(a)$ are covered by Eq.(2.5.9) with $M = 1$. Hence

$$I(x) = (\mu-1)!f(a)\frac{e^{-iax}}{(ix)^{\mu}} - \frac{e^{-ibx}}{ix}(b-a)^{\mu-1}f(b) + \frac{\unicode{x043A}}{x} \qquad (2.5.10)$$

when f' is absolutely integrable.

The formula of Eq.(2.5.10) was arrived at by constructing a function which vanished at $t = a$ and was continuous from a to ∞ so that Theorem 2.5.2 could be invoked. When f'' is absolutely integrable we can try to arrange that the same property is available for the derivative. To this end let

$$f_1(t) = f(a) + (t-a)f'(a)$$

and

$$f_2(t) = (t-a)^{\mu-1}\{f(t) - f_1(t)\},$$
$$f_3(t) = f_2(b) + (t-b)f_2'(b).$$

Write the integral as

$$\int_a^b (t-a)^{\mu-1}\{f(t) - f_1(t)\}e^{-(\epsilon+ix)t}dt + \int_b^\infty f_3(t)e^{-(\epsilon+ix)t}dt$$
$$+ \int_a^b (t-a)^{\mu-1}f_1(t)e^{-(\epsilon+ix)t}dt - \int_b^\infty f_3(t)e^{-(\epsilon+ix)t}dt.$$

The combination of the first two integrals has an integrand which, together with its first derivative, is continuous from a to ∞ and vanishes at $t = a$. Therefore, Theorem 2.5.2 asserts that there will be a contribution of $\varkappa/x^2$. In the other two integrals Eq.(2.5.9) can be quoted with $M = 2$ for the terms in $f(a)$ and $f'(a)$ (with μ increased to $\mu+1$). Accordingly

$$I(x) = e^{-iax}\left\{\frac{(\mu-1)!}{(ix)^\mu}f(a) + \frac{\mu!}{(ix)^{\mu+1}}f'(a)\right\}$$
$$-e^{-ibx}\left\{\frac{(b-a)^{\mu-1}}{ix}f(b) + \frac{1}{(ix)^2}\frac{d}{db}(b-a)^{\mu-1}f(b)\right\} + \frac{\varkappa}{x^2}.$$

Clearly, if more derivatives of f are integrable, the same technique can be deployed by adding further terms in the Taylor expansions defining f_1 and f_3. The net result is

$$I(x) = e^{-iax}\sum_{p=0}^{N-1}\frac{(\mu+p-1)!}{p!(ix)^{\mu+p}}f^{(p)}(a) - e^{-ibx}\sum_{p=0}^{N-1}\frac{1}{(ix)^{p+1}}\frac{d^p}{db^p}(b-a)^{\mu-1}f(b) + \frac{\varkappa}{x^N}$$

(2.5.11)

when $f, f', \ldots, f^{(N)}$ are absolutely integrable over (a, b).

When the singularity is at the upper limit e.g.

$$I_1(x) = \int_a^b (b-t)^{\nu-1}f(t)e^{-ixt}dt$$

where $0 < \nu \le 1$, switching t to $-t$ converts the integral to one of the type just considered. Hence

$$I_1(x) = e^{-ibx}\sum_{p=0}^{N-1}\frac{(\nu+p-1)!}{p!(-ix)^{\nu+p}}(-)^p f^{(p)}(b)$$
$$+ e^{-iax}\sum_{p=0}^{N-1}\frac{1}{(ix)^{p+1}}\frac{d^p}{da^p}\{(b-a)^{\nu-1}f(a)\} + \frac{\varkappa}{x^N}. \qquad (2.5.12)$$

Example 2.5.3 With the assistance of Example 2.5.2 the more general

$$\int_a^b (t-a)^{\mu-1}(b-t)^{\nu-1}f(t)e^{-ixt}dt$$

can be tackled. Select c as the mid-point of the interval (a, b). For the integral from a to c call on Eq.(2.5.11) but with $(b-t)^{\nu-1}f(t)$ in place of $f(t)$. The integral from c to b can be dealt with by Eq.(2.5.12) on replacing $f(t)$ by $(t-a)^{\mu-1}f(t)$. It turns out that the two contributions from c cancel so that

$$\int_a^b (t-a)^{\mu-1}(b-t)^{\nu-1}f(t)e^{-ixt}dt$$

$$= e^{-iax}\sum_{p=0}^{N-1}\frac{(\mu+p-1)!}{p!(ix)^{\mu+p}}\frac{d^p}{da^p}\{(b-a)^{\nu-1}f(a)\}$$

$$+e^{-ibx}\sum_{p=0}^{N-1}\frac{(\nu+p-1)!}{p!(-ix)^{\nu+p}}(-)^p\frac{d^p}{db^p}\{(b-a)^{\mu-1}f(b)\}+\frac{\varkappa}{x^N}. \quad (2.5.13)$$

In these examples the behaviour for unlimited x comes from the endpoints of the interval of integration. For the general interval there may be interior points where f has an integrable singularity or where the continuity of f or one of its derivatives fails. It is then necessary to split the range of integration at such points so that the integrals over the sub-intervals are of the type already considered. However, the continuity of the derivatives does not have to be examined beyond the order of the error which is acceptable. For instance, if $\varkappa/x^N$ is acceptable, there is no point in looking at the continuity of the $(N+1)$th derivative or one of higher order.

2.6   Method of stationary phase

A generalisation of the Fourier integral is

$$\int_a^b f(t)e^{-ixh(t)}dt$$

where a, b are real $(b > a)$ and $h(t)$ is real-valued. The obvious mode of attack is to put $h(t) = u$ and revert to the theory of the preceding section. According to that, contributions can be expected from $t = a$ and $t = b$ when the integrand is smooth enough throughout the range of integration. The integrand involves $h'(t)$ after the proposed substitution so that the behaviour of h' is also a relevant factor. Indeed, it has been seen in Section 2.3 that places where $h'(t) = 0$ can produce important contributions.

A rough argument is that, away from a point where $h'(t) = 0$, the exponential oscillates very fast while $f(t)$ scarcely changes. The consequent cancellation results in a small contribution. On the other hand, near a point where $h'(t) = 0$ the oscillations are slower and a possible contribution may ensue. A good estimate for it may be obtained from approximations to f and h in the neigbourhood. Such an estimate is said to be attained by the *method of stationary phase*.

Although this rough argument is plausible it is hard to justify. So we shall proceed differently and, at the same time, allow for more variations in h' than at a simple stationary point. Points where h' vanishes or has other singular behaviour are known often as *critical points*. It is essential to realise that the contribution of a critical point to a general integral may not be dominant. The contributions of the endpoints may be more significant as may be those from singularities of f. All these contributions must be taken into account before deciding that the principal approximation to an integral stems from a critical point. Having issued that warning we shall assume that the range of integration has been split into sub-intervals so that no sub-interval contains more than one critical point which coincides with an endpoint.

In view of these assumptions consider

$$\int_a^b f(t)e^{-ixh(t)}dt$$

where $h'(t) > 0$ for $t > a$ but $h'(t) \simeq a_0(t-a)^{\mu-1}$ with $\mu > 0$ for $t \simeq a$. Make the change of variable $h(t) - h(a) = u$; then the integral becomes

$$e^{-ixh(a)} \int_0^B \frac{f(t)}{h'(t)} e^{-ixu} du$$

where $B = h(b) - h(a)$. If $f(t) \simeq b_0(t-a)^{\lambda-1}$ with $\lambda > 0$ for $t \simeq a$

$$\frac{f(t)}{h'(t)} \approx \frac{b_0}{a_0}(t-a)^{\lambda-\mu} \approx \frac{b_0}{a_0}\left(\frac{\mu u}{a_0}\right)^{\frac{\lambda-1}{\mu}}$$

for u near the origin. Assume that the derivative of $f(t)/h'(t)$ with respect to u is absolutely integrable from 0 to B. Then Eq.(2.5.10) may be cited and

$$\int_a^b f(t)e^{-ixh(t)}dt = \left(\frac{\lambda}{\mu}-1\right)!\frac{b_0}{\mu}\left(\frac{\mu}{ia_0x}\right)^{\frac{\lambda}{\mu}}e^{-ixh(a)} + \frac{e^{-ixh(b)}}{ix}\frac{f(b)}{h'(b)} + \frac{\upsilon}{x}. \quad (2.6.1)$$

The first term on the right of Eq.(2.6.1) constitutes the contribution of the critical point whereas the second term is due to an endpoint. If $\lambda < \mu$ the critical point provides the dominant performance; in particular, this is true for a point of stationary phase with $\mu = 2$ so long as $\lambda < 2$. For $\lambda > \mu$ the endpoint dominates and, in fact, the contribution of the critical point could be absorbed in the term υ/x.

Further assumptions about f and h' may permit the calculation of more terms through Eq.(2.5.11). The formulae do tend to become rather cumbersome and, in a practical application, it may be more economical to proceed directly than to try to insert values into general expressions.

The case $h'(t) > 0$ for $t > a$ has been discussed. When $h'(t) < 0$ for $t > a$ note that $ixh(t) = -ixh_1(t)$ where $h_1(t) = -h(t)$. Since $h_1'(t) > 0$ a return to the previous case has been achieved but with $-ix$ in place of ix. Then Eq.(2.6.1) may be quoted in terms of h_1 and $-ix$. After going back to h, Eq.(2.6.1) is recovered unchanged but do not forget how the phase of quantities like ia_0x was defined after Theorem 2.5.2.

When the critical point is at b instead of a it is necessary to specify h and f near b. Assume that, for $t \simeq b$, $h'(t) \simeq a_0(b-t)^{\mu-1}$, $f(t) \simeq b_0(b-t)^{\lambda-1}$. The change of variable from t to $-t$ puts the critical point at the lower end of integration and Eq.(2.6.1) is available again. In this case

$$\int_a^b f(t)e^{-ixh(t)}dt = \left(\frac{\lambda}{\mu}-1\right)!\frac{b_0}{\mu}\left(\frac{\mu}{ia_0x}\right)^{\frac{\lambda}{\mu}}e^{-ixh(b)} + \frac{e^{-ixh(a)}}{ix}\frac{f(a)}{h'(a)} + \frac{\mathsf{u}}{x}. \quad (2.6.2)$$

Example 2.6.1 In this example h has a single critical point at $t = c$ lying between a and b. It is assumed that $\mu = 2$ and that f is non-singular. For the integral from a to c employ Eq.(2.6.2) with $b = c$. Express the integral from c to b by Eq.(2.6.1) with $a = c$. Since $\lambda = 1$, $b_0 = f(c)$ and $a_0 = h''(c)$. Hence

$$\int_a^b f(t)e^{-ixh(t)}dt = f(c)\left\{\frac{2\pi}{ih''(c)x}\right\}^{\frac{1}{2}} + O\left(\frac{1}{x}\right). \quad (2.6.3)$$

2.7   Expansions

For many of the integrals in this chapter the approximations have not extended beyond one term. Often this is sufficient for most purposes. The reason is that most of the estimates contain an infinitesimal factor and so they tend to be used for error corrections in asymptotic expansions. For example, in Laplace integrals we try to write

$$f(t) = \sum_{n=0}^{N} f_n(t) + R_N(t)$$

where the integral of f_n can be evaluated exactly. The hope is that it will be possible to show that the effect of R_N is small. When the integral of f_n cannot be calculated exactly the situation is more complex because it has to be shown that the errors do not combine into an appreciable quantity.

Details on the integration of asymptotic series will be found in the next chapter.

Exercises on Chapter 2

1. Prove that

$$\int_0^\infty \frac{e^{-xt}}{(1+t^2)^{\frac{1}{2}}}dt = (1 + и)/x.$$

2. By putting $t = xu$ show that

$$\int_0^\infty e^{x\ln t - t}\ln t\, dt = (2\pi x)^{\frac{1}{2}}(1 + и)e^{x\ln x - x}\ln x.$$

3. Prove that

$$\int_0^{\pi/2} te^{z\cos t}dt = e^z(1 + и)/z$$

for $|\mathrm{ph}\, z| \leq \pi/2 - \delta$.

4. Prove that, for $\mu > 0$, with analytic continuation

$$\int_0^\infty e^{-zt}(1+t)^{\mu-1}dt = (1 + и)/z$$

for $|\mathrm{ph}\, z| \leq 3\pi/2 - \delta$.

5. Obtain an approximation for $\int_{1/2}^\infty \exp\{-x(w - \ln w)\}dw$.

6. If C is a contour which starts at $w = \infty e^{-\pi i/3}$ and ends at $w = \infty e^{\pi i/3}$ show that

$$\int_C \exp\{x(\tfrac{1}{3}w^3 - w)\}dw = \frac{\pi^{\frac{1}{2}}i}{x^{\frac{1}{2}}}(1 + и)e^{-2x/3}.$$

7. By taking a derivative with respect to μ of the equation before Eq.(2.5.4) show that

$$\lim_{\epsilon \to +0}\int_a^\infty (t-a)^{\mu-1}\ln(t-a)e^{-(\epsilon+ix)t}dt = \frac{(\mu-1)!}{(ix)^\mu}e^{-ixa}\{\psi(\mu-1) - \ln|x| - \frac{i\pi}{2}\mathrm{sgn}\,x\}$$

where $\psi(z) = z!'/z!$. This is exactly the same as would be obtained by taking a derivative with respect to μ of the right-hand side of Eq.(2.5.4). Use this fact and Eq.(2.5.13) to show that

$$\int_a^b (t-a)^{\mu-1}(b-t)^{\nu-1}\ln(t-a)f(t)e^{-ixt}dt$$

$$= e^{-iax}\sum_{p=0}^{N-1}\frac{(\mu+p-1)!}{p!(ix)^{\mu+p}}$$

$$\times \left\{\psi(\mu+p-1) - \ln|x| - \frac{i\pi}{2}\mathrm{sgn}\,x\right\}\frac{d^p}{da^p}\{(b-a)^{\nu-1}f(a)\}$$

$$+ e^{-ibx}\sum_{p=0}^{N-1}\frac{(\nu+p-1)!}{p!(-ix)^{\nu+p}}(-)^p\frac{d^p}{db^p}\{(b-a)^{\mu-1}f(b)\ln(b-a)\} + \frac{и}{x^N}.$$

8. How do $\int_{-\infty}^\infty e^{-t^2-itx}dt$ and $\int_0^\infty e^{-t^2-itx}dt - \int_{-\infty}^0 e^{-t^2-itx}dt$ differ as $x \to \infty$?

9. Show that

$$\int_0^1 e^{-ixt} \ln(1+t)dt = -\frac{e^{-ix}}{ix} \ln 2 + \sum_{p=1}^{N-1} \frac{\{p-1\}!}{(ix)^{p+1}} (-)^{p+1} \left(1 - \frac{e^{-ix}}{2^p}\right) + \frac{и}{x^N}.$$

10. Prove that

$$\int_0^1 e^{-ixt^3}dt = \frac{1}{3}! \left(\frac{1}{ix}\right)^{1/3} - \frac{e^{-ix}}{3ix} + \frac{и}{x}.$$

11. Show that

$$\int_0^{\pi/2} te^{-ix\cos t}dt = \frac{1}{ix} \left(\frac{\pi}{2} - e^{-ix}\right) + \frac{и}{x}.$$

and deduce that

$$\int_0^{\pi/2} t\sin(x\cos t)dt = \frac{1}{x} \left(\frac{\pi}{2} - \cos x\right) + \frac{и}{x}.$$

Chapter 3
SERIES

Chapter 2 was concerned primarily with approximations to integrals in which relatively few assumptions were made about the form of the integrand. When more is known about the integrand more information can be obtained. We will consider here the possibility that $f(z)$ can be represented by a series which may be asymptotic. Recall, from Chapter 1, that $f(z) \sim \sum_{m=1}^{\infty} c_m \varphi_m(z)$ as $z \to 0$ if $f(z) = \sum_{m=1}^{n} c_m \varphi_m(z) + R_n(z)$ and $R_n(z) = (c_{n+1} + \varkappa)\varphi_{n+1}(z)$ for every standard n.

3.1 General results

The first case to be discussed is where

$$f(z) \sim \sum_{m=1}^{\infty} c_m z^{\lambda_m}$$

as $z \to 0$ with ph $z = \alpha$, the coefficients c_m being standard. The constants λ_m satisfy

$$-1 < \mathcal{R}\lambda_1 < \mathcal{R}\lambda_2 < \mathcal{R}\lambda_3 < \cdots$$

so that the sequence $\{z^{\lambda_m}\}$ is asymptotic. Here z^{λ_m} means $|z|^{\lambda_m} \exp(i\lambda_m \alpha)$.

Theorem 3.1.1 *If $|f(re^{i\alpha})| < Ae^{\gamma r}$ for $r > 0$, with A and γ standard,*

$$\int_0^{\infty} e^{-t} f(zt) dt \sim \sum_{m=1}^{\infty} \lambda_m! c_m z^{\lambda_m}$$

as $z \to 0$ with ph $z = \alpha$.

Proof. The existence of the integral is assured because the growth of $f(zt)$ is limited by $\exp(|z| \gamma t)$ and $\gamma |z|$ is infinitesimal. Also

$$\int_0^{\infty} e^{-t} f(zt) dt = \sum_{m=1}^{n} \lambda_m! c_m z^{\lambda_m} + \int_0^{\infty} e^{-t} R_n(zt) dt$$

for every standard n. Now $R_n(t)/t^{\lambda_{n+1}}$ is bounded for infinitesimal t by the definition of an asymptotic expansion and exponentially bounded for larger

43

t because f is. Hence, by writing $R_n(t) = t^{\lambda_{n+1}} R_n(t)/t^{\lambda_{n+1}}$, we can call on Theorem 2.1.1 of Chapter 2 to confirm that

$$\int_0^\infty e^{-t} R_n(zt) dt = \lambda_{n+1}! z^{\lambda_{n+1}}(c_{n+1} + \varkappa)(1 + \varkappa) = \lambda_{n+1}! z^{\lambda_{n+1}}(c_{n+1} + \varkappa).$$

Thus, the remainder behaves as it should for the asymptotic nature of the expansion and the theorem is established. ∎

A rather more general theorem is

Theorem 3.1.2 *Under the conditions of Theorem 3.1.1*

$$\int_0^{\infty e^{i\beta}} e^{-t} f(zt) dt \sim \sum_{m=1}^\infty \lambda_m! c_m z^{\lambda_m}$$

as $z \to 0$ with ph $z = \alpha - \beta$ *and* $|\beta| \le \pi/2 - \delta$.

Proof. A different method of showing that the remainder has the correct form has to be adopted because Theorem 2.1.1 of Chapter 2 is not immediately available. When t is limited $zt \simeq 0$ and

$$R_n(zt)/z^{\lambda_{n+1}} \simeq c_{n+1} t^{\lambda_{n+1}}; \qquad (3.1.1)$$

in addition

$$\left| e^{-t} R_n(zt)/z^{\lambda_{n+1}} \right| \le (1 + |c_{n+1}|) A_n |t|^{\lambda_{n+1}} \exp(-|t| \cos \beta) \qquad (3.1.2)$$

for some limited A_n. When zt is not infinitesimal, $|t|$ must be unlimited and so $e^{-t/2}/z^{\lambda_{n+1}}$ is infinitesimal. Hence, since n is limited,

$$
\begin{aligned}
\left| e^{-t} R_n(zt)/z^{\lambda_{n+1}} \right| &= \left| \frac{e^{-t/2}}{z^{\lambda_{n+1}}} \cdot e^{-t/2} \left\{ f(zt) - \sum_{m=1}^n c_m z^{\lambda_m} t^m \right\} \right| \\
&\le A_n' \exp(-\tfrac{1}{2} |t| \cos \beta + \gamma' |zt|) \\
&\le A_n' \exp(-\tfrac{1}{4} |t| \cos \beta) \qquad (3.1.3)
\end{aligned}
$$

for some standard A_n' and γ'. It follows from Eq.(3.1.2) and Eq.(3.1.3) that $e^{-t} R_n(zt)/z^{\lambda_{n+1}}$ is bounded by an absolutely integrable function which is also a bound for $c_{n+1} e^{-t} t^{\lambda_{n+1}}$. Therefore, Eq.(3.1.1) and Theorem A.5.5 supply

$$
\begin{aligned}
\int_0^{\infty e^{i\beta}} e^{-t} R_n(zt) dt/z^{\lambda_{n+1}} &\simeq \int_0^{\infty e^{i\beta}} c_{n+1} e^{-t} t^{\lambda_{n+1}} dt \\
&= \lambda_{n+1}! c_{n+1} + \varkappa
\end{aligned}
$$

and the theorem is proved. ∎

A variant for the Laplace integral is

Theorem 3.1.3 (Watson's Lemma) *Under the conditions of Theorem 3.1.1*

$$\int_0^{\infty e^{i\alpha}} e^{-zt} f(t) \sim \sum_{m=1}^\infty \lambda_m! c_m / z^{1+\lambda_m}$$

for unlimited z with $|\alpha + \text{ph}\, z| \leq \frac{1}{2}\pi - \delta$, δ being standard and positive.
Proof. The proof is the same as for Theorem 3.1.1 but deploying Corollary 2.2.4 of Chapter 2 in checking the behaviour of the remainder. ∎

The proof of Theorem 3.1.2 is virtually unaltered by the insertion of the factor t^μ or $t^\mu \ln t$ with $\mathcal{R}\mu > 0$. Thus we have

Theorem 3.1.4 *Under the conditions of Theorem 3.1.1, as $z \to 0$ with $\text{ph}\, z = \alpha - \beta$ and $|\beta| \leq \pi/2 - \delta$*

$$\int_0^{\infty e^{i\beta}} e^{-t} t^\mu f(zt)\, dt \;\sim\; \sum_{m=1}^\infty (\lambda_m + \mu)!\, c_m z^{\lambda_m},$$

$$\int_0^{\infty e^{i\beta}} e^{-t} t^\mu \ln t\, f(zt)\, dt \;\sim\; \sum_{m=1}^\infty (\lambda_m + \mu)!\, \psi(\lambda_m + \mu) c_m z^{\lambda_m}$$

where $\mathcal{R}\mu \geq 0$ and $\psi(z) = z!'/z!$.
Similarly, we have generalisations of Watson's lemma

Theorem 3.1.5 *For unlimited z with $|\alpha + \text{ph}\, z| \leq \frac{1}{2}\pi - \delta$*

$$\int_0^{\infty e^{i\alpha}} e^{-zt} t^\mu f(t)\, dt \;\sim\; \sum_{m=1}^\infty (\lambda_m + \mu)!\, c_m / z^{1+\lambda_m+\mu},$$

$$\int_0^{\infty e^{i\alpha}} e^{-zt} t^\mu \ln t\, f(t)\, dt \;\sim\; \sum_{m=1}^\infty (\lambda_m + \mu)!\, c_m \{\psi(\lambda_m + \mu)$$
$$- \ln|z| - i\,\text{ph}\, z\} / z^{1+\lambda_m+\mu}.$$

Example 3.1.1 *The Airy function*
The Airy function $\text{Ai}(z)$ can be defined by

$$\text{Ai}(z) = \frac{1}{2\pi i} \int_{\infty e^{-2\pi i/3}}^{\infty e^{2\pi i/3}} e^{-(\frac{1}{3}t^3 - zt)}\, dt \tag{3.1.4}$$

where the contour of integration is the solid line of Fig. 3.1.1. The integral converges for any z. The Airy function is an entire function of z and has the series representation

$$\text{Ai}(z) \;=\; \frac{1}{(-\frac{1}{3})!3^{\frac{2}{3}}}\left(1 + \frac{z^3}{3!} + \frac{1.4}{6!}z^6 + \frac{1.4.7}{9!}z^9 + \cdots\right)$$
$$- \frac{1}{(-\frac{2}{3})!3^{\frac{1}{3}}}\left(z + \frac{2}{4!}z^4 + \frac{2.5}{7!}z^7 + \frac{2.5.8}{10!}z^{10} + \cdots\right). \tag{3.1.5}$$

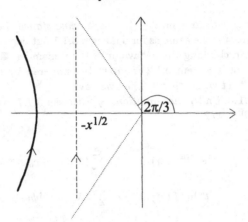

Figure 3.1.1 Contour for the Airy function

The asymptotic behaviour of the two series in Eq.(3.1.5) is discussed later in Section 3.4 where it is shown that the dominant terms of each cancel. Therefore, the definition of Eq.(3.1.5) is not pertinent for settling how $\mathrm{Ai}(z)$ behaves as $|z| \to \infty$ although the corresponding one for $\mathrm{Bi}(z)$ is relevant.

A representation suitable for elucidating the conduct of $\mathrm{Ai}(z)$ for unlimited z is

$$\mathrm{Ai}(z) = \frac{1}{2\pi} \exp(-\tfrac{2}{3}z^{\frac{3}{2}}) \int_0^\infty \exp(-z^{\frac{1}{2}}u) \cos \tfrac{1}{3}u^{\frac{3}{2}} \frac{du}{u^{\frac{1}{2}}} \qquad (3.1.6)$$

when $|\mathrm{ph}\, z| < \pi$. In Eq.(3.1.6) z^β means $|z|^\beta \exp(i\beta\,\mathrm{ph}\, z)$. To verify Eq.(3.1.6) put $z = x$ where x is positive. The contour in Eq.(3.1.4) can be deformed then into the line parallel to the imaginary axis through $-x^{\frac{1}{2}}$, shown dashed in Fig. 3.1.1. On the part above the real axis put $t = -x^{\frac{1}{2}} + u^{\frac{1}{2}}i$ and on the part below $t = -x^{\frac{1}{2}} - u^{\frac{1}{2}}i$. Eq.(3.1.6) is recovered with $z = x$. But Eq.(3.1.6) supplies a regular function of z for $|\mathrm{ph}\, z| < \pi$ and so Eq.(3.1.6) is valid, subject to the restriction on phase, by analytic continuation.

The convergent expansion

$$\cos \tfrac{1}{3}u^{\frac{3}{2}} = \sum_{m=0}^\infty \frac{(-)^m}{(2m)!} (\tfrac{1}{3}u^{\frac{3}{2}})^{2m}$$

permits the application of Theorem 3.1.5 with $\lambda_m = 3m$ and $\mu = -\tfrac{1}{2}$. Hence, for unlimited z,

$$\mathrm{Ai}\,(z) \sim \frac{1}{2\pi z^{\frac{1}{4}}} \exp\left(-\tfrac{2}{3}z^{\frac{3}{2}}\right) \sum_{m=0}^\infty \frac{(3m - \tfrac{1}{3})!(-)^m}{(2m)!3^{2m}z^{3m/2}} \qquad (3.1.7)$$

when $|\mathrm{ph}\, z| < \pi$.

To gain a formula when ph z passes through π without having recourse to another integral representation we note that Ai(z) is a solution of

$$\frac{d^2y}{dz^2} = zy. \tag{3.1.8}$$

Confirmation is obtained easily from Eq.(3.1.4) because of the excellent convergence of the integral. Observe that, if $\eta = e^{2\pi i/3}$, Eq.(3.1.8) is unaltered by changing z to ηz or to $\eta^2 z$. Therefore Ai(ηz) and Ai($\eta^2 z$) are solutions of Eq.(3.1.8). The theory of differential equations asserts that only two solutions are linearly independent. Hence there must be constants A and B such that

$$\text{Ai}(z) + A\,\text{Ai}(\eta z) + B\,\text{Ai}(\eta^2 z) = 0$$

for all z. Put $z = 0$ with the result

$$1 + A + B = 0$$

since Ai(0) $\neq 0$ from Eq.(3.1.5). Take a derivative with respect to z and put $z = 0$; then

$$1 + \eta A + \eta^2 B = 0.$$

We conclude that $A = \eta$ and $B = \eta^2$ so that

$$\text{Ai}(z) + \eta\,\text{Ai}(\eta z) + \eta^2\,\text{Ai}(\eta^2 z) = 0. \tag{3.1.9}$$

Suppose now that $\pi/3 < \text{ph}\,z < 5\pi/3$. Since Ai($z$) is entire the phase of its argument can be altered by 2π without any effect. Take η to be $e^{-4\pi i/3}$ in Eq.(3.1.9); then $-\pi < \text{ph}(\eta z) < \pi/3$. Similarly, with $\eta^2 = e^{-2\pi i/3}$, the phase satisfies $-\pi/3 < \text{ph}(\eta^2 z) < \pi$. Therefore Eq.(3.1.7) is available for the second and third terms of Eq.(3.1.9); accordingly

$$\begin{aligned}
\text{Ai}(z) \;\sim\; & \frac{1}{2\pi z^{\frac{1}{4}}} \exp\left(-\tfrac{2}{3}z^{\frac{3}{2}}\right) \sum_{m=0}^{\infty} \frac{(3m - \frac{1}{2})!(-)^m}{(2m)!3^{2m}z^{3m/2}} \\
& + \frac{i}{2\pi z^{\frac{1}{4}}} \exp\left(\tfrac{2}{3}z^{\frac{3}{2}}\right) \sum_{m=0}^{\infty} \frac{(3m - \frac{1}{2})!}{(2m)!3^{2m}z^{3m/2}}
\end{aligned} \tag{3.1.10}$$

for $\pi/3 < \text{ph}\,z < 5\pi/3$. Notice that, as ph z increases from $\pi/3$, the second term of Eq.(3.1.10) is exponentially small compared with the first term until ph z approaches π.

A yet more general version of Theorem 3.1.4 can be derived by defining $f_1(z) = f(z^{1/\nu})$ with $\nu \geq 1$. Then, according to Theorem 3.1.4,

$$\int_0^{\infty e^{i\beta}} e^{-t} t^{\frac{\mu+1}{\nu}-1} f_1(z^\nu t)\,dt \sim \sum_{m=1}^{\infty} \left(\frac{\lambda_m + \mu + 1}{\nu} - 1\right)!\,c_m z^{\lambda_m}.$$

Now put $t = u^\nu$ to obtain

Theorem 3.1.6 *Under the conditions of Theorem 3.1.1*

$$\int_0^{\infty e^{i\beta/\nu}} e^{-u^\nu} u^\mu f(zu) du \sim \frac{1}{\nu} \sum_{m=1}^{\infty} \left(\frac{\lambda_m + \mu + 1}{\nu} - 1 \right)! c_m z^{\lambda_m}$$

as $z \to 0$ with ph $z = \alpha - \beta/\nu$ *and* $|\beta| \leq \pi/2 - \delta$.

The presence of u^ν in the exponent of Theorem 3.1.6 means that the integral could exist for more rapid growth of f than is permitted by Theorem 3.1.1. However, any extension of Theorem 3.1.6 which could be reached by taking advantage of this fact will be left on one side.

It will be noticed that the effect of integration is to multiply the coefficients by factorial factors which grow with m and yet the expansion remains asymptotic. The next theorem shows that even larger factors can be inserted without destroying the asymptotic nature provided that the growth of f is not too large.

Theorem 3.1.7 *If $|f(re^{i\alpha})| < Ar^p + B$ for $r \geq 0$, with A, B and p standard constants*

$$\int_{-\infty+i\beta}^{\infty+i\beta} e^{-t^2/2} f(ze^t) dt \sim \sum_{m=1}^{\infty} (2\pi)^{\frac{1}{2}} c_m e^{\lambda_m^2/2} z^{\lambda_m}$$

as $z \to 0$ with ph $z = \alpha - \beta$ *and β limited.*

Proof. For a typical term in the asymptotic expansion of f Cauchy's theorem enables the contour being pushed on to the real axis because the t^2 ensures that there is no contribution from pieces at $\pm\infty$. Thus

$$\int_{-\infty+i\beta}^{\infty+i\beta} e^{\lambda_m t - t^2/2} dt = (2\pi)^{\frac{1}{2}} e^{\lambda_m^2/2}.$$

Consequently, all that has to be done is to check that the remainder behaves appropriately.

Now

$$\int_{-\infty+i\beta}^{\infty+i\beta} e^{-t^2/2} R_n(ze^t) dt = \int_{-\infty}^{\infty} e^{-(u+i\beta)^2/2} R_n(|z| e^{i\alpha+u}) du.$$

When u is limited so is e^u and $|z| e^u \simeq 0$. So, for limited u,

$$R_n(|z| e^{i\alpha+u})/z^{\lambda_{n+1}} \simeq c_{n+1} e^{\lambda_{n+1}(u+i\beta)}.$$

The desired result will follow from Theorem A.5.5 so long as the integrand involving R_n is bounded for all u by an absolutely integrable function.

If u is such that ze^u is infinitesimal

$$\left| e^{-(u+i\beta)^2/2} R_n(|z| e^{i\alpha+u})/z^{\lambda_{n+1}} \right| < A_n e^{\frac{1}{2}(\beta^2 - u^2) + \mathcal{R}\lambda_{n+1}u}(1 + |c_{n+1}|)$$

for some standard A_n. If ze^u is not infinitesimal so that u is not limited

$$
\left| e^{-(u+i\beta)^2/2} R_n(|z|\, e^{i\alpha+u})/z^{\lambda_{n+1}} \right|
$$

$$
= \left| \frac{e^{-(u+i\beta)^2/2}}{z^{\lambda_{n+1}}} \left\{ f(|z|\, e^{i\alpha+u}) - \sum_{m=1}^{n} c_m \left(|z|\, e^{i\alpha+u} \right)^{\lambda_m} \right\} \right|
$$

$$
< \frac{e^{\frac{1}{2}(\beta^2-u^2)}}{z^{\lambda_{n+1}}} \left\{ A\,|z|^p\, e^{pu} + B + \sum_{m=1}^{n} \left| c_m \left(|z|\, e^{i\alpha+u} \right)^{\lambda_m} \right| \right\}
$$

$$
< A' e^{-u^2/8 + \lambda u}
$$

where λ is standard and exceeds $\mathcal{R}\lambda_{n+1} + \max(p, \mathcal{R}\lambda_n)$. Absolute integrability has been confirmed and the proof is complete. ■

One can envisage constructing other asymptotic expansions by means of various integrals. This possibility will not be explored further because, when the coefficients are as large as in Theorem 3.1.7, one wonders whether they have practical value. Asymptotic expansions with such large coefficients are not expected to converge and have to be truncated in a practical computation. The value of the truncated series as an approximation to the integral is determined by the size of the error caused by the truncation. In other words, an estimate of the remainder after truncation is an essential feature in settling the value of an asymptotic expansion. A start on this topic is made in the next section.

3.2   Power series

Suppose that

$$
g(w) = \sum_{m=0}^{\infty} b_m w^m \tag{3.2.1}
$$

where w is a complex variable. Assume that the series for g has a non-zero, but finite, radius of convergence c. Put $w = cz$ and $g(cz) = f(z)$; then

$$
f(z) = \sum_{m=0}^{\infty} a_m z^m \tag{3.2.2}
$$

where $a_m = c^m b_m$. The series for f has a radius of convergence of unity which is a convenient normalisation. So from now on we shall treat Eq.(3.2.2) leaving it to the reader to make any necessary conversions to Eq.(3.2.1).

As regards the coefficients in Eq.(3.2.2) it will be assumed that $\{a_n\}$ is a standard sequence i.e. a_n is a standard constant for standard n. A sufficient condition for a unit radius of convergence is

$$
\lim_{m\to\infty} |a_{m+1}/a_m| = 1. \tag{3.2.3}
$$

For our purposes a little more information than is contained in Eq.(3.2.3) is required and we shall assume that

$$a_{m+1}/a_m = 1 + (a + \varkappa)/m^{\frac{1}{2}} \tag{3.2.4}$$

for all unlimited m, a being a standard constant. The condition of Eq.(3.2.4) is consistent with Eq.(3.2.3) (see Theorem A.2.2) and can be stated in the alternative fashion:

for every standard $\epsilon > 0$ there is a standard $N(\epsilon)$ such that

$$\left| \frac{a_{m+1}}{a_m} - 1 - \frac{a}{m^{\frac{1}{2}}} \right| < \frac{\epsilon}{m^{\frac{1}{2}}} \tag{3.2.5}$$

for $m \geq N(\epsilon)$.

Associated with the f of Eq.(3.2.2) is

$$\varphi(z) = \sum_{m=0}^{\infty} \frac{a_m}{m!} z^m. \tag{3.2.6}$$

The series for φ converges for all z and so is an entire function. Roughly speaking, all the a_m are the same for unlimited m according to Eq.(3.2.4) so that $\varphi(z)$ behaves something like e^z to a first approximation. In fact, a bound for φ is furnished by the following theorem.

Theorem 3.2.1 *Under the condition of Eq.(3.2.4) there are standard A and γ such that $|\varphi(z)| \leq A \exp(\gamma |z|)$.*

Proof. Only large $|z|$ needs to be considered. Choose ϵ positive and standard. Then, with N as for Eq.(3.2.5),

$$\left| \sum_{m=0}^{N-1} \frac{a_m}{m!} z^m \right| \leq |z|^N \sum_{m=0}^{N-1} \frac{|a_m|}{m!} \leq A' \exp(\gamma' |z|)$$

for some standard A' and γ'. By virtue of Eq.(3.2.5) there is a standard γ'' such that $|a_{m+1}/a_m| < \gamma''$ for $m \geq N$. Hence

$$\left| \sum_{m=N}^{\infty} \frac{a_m}{m!} z^m \right| \leq |a_N| \sum_{m=N}^{\infty} \frac{(\gamma'' |z|)^m}{m!} \leq |a_N| \exp(\gamma'' |z|).$$

Since a_N is standard, addition of the two inequalities supplies the theorem. ■

Theorem 3.2.1 shows that $\varphi(z)$ is an entire function of exponential type when subject to Eq.(3.2.4). Later on more precise estimates will be secured. In the meantime, Theorem 3.2.1 will suffice. For instance, it shows that the Laplace transform of Theorem 3.1.3 of $\varphi(t)$ exists for large enough $|z|$. Indeed, Theorem 3.1.3 indicates that it is asymptotically $f(1/z)/z$; the convergence of Eq.(3.2.2) ensures that the Laplace transform is $f(1/z)/z$ for $|z| > 1$.

Consider

$$F(z) = \int_0^\infty e^{-t} \varphi(zt) dt. \qquad (3.2.7)$$

From Theorems 3.1.1 and 3.2.1, $F(z) \sim f(z)$ as $z \to 0$ which implies that $F(z) = f(z)$ wherever the series for $f(z)$ converges i.e. in the unit circle. However, $F(z)$ may exist for z outside the unit circle and offers a route for analytic continuation of $f(z)$. Suppose that

$$\left| \varphi(r_0 e^{i\alpha} t) \right| < K e^{Mt} \qquad (3.2.8)$$

where K is independent of r_0 and t while $M < 1$. The integral in Eq.(3.2.7) with $z = r_0 e^{i\alpha}$ is convergent then. Moreover, if $r < r_0$,

$$\left| \varphi(r e^{i\alpha} t) \right| < K \exp(Mrt/r_0)$$

so that the integral converges for $z = r e^{i\alpha}$. When $r_0 > 1$, $F(r e^{i\alpha})$ converges all the way from the origin to a point outside the unit circle and hence provides an analytic continuation of $f(z)$ across the arc of the unit circle near $e^{i\alpha}$.

This result enables us to say something about the location of the singularities of $f(z)$ and its analytic continuation. Firstly, $f(z)$ has no singularities in $|z| < 1$ which entails Eq.(3.2.8) being true for any α and $r_0 < 1$. Consequently, the γ in Theorem 3.2.1 could be brought down to 1. Secondly, if

$$\left| \varphi(e^{i\alpha} t) \right| < K e^{Mt}$$

with $M < 1$, $f(z)$ can be continued across the arc at $e^{i\alpha}$ and any singularity on the ray is no closer than $e^{i\alpha}/M$. An application to $\varphi(z) = e^z$ which satisfies $|\varphi(zt)| < \exp(tr \cos \alpha)$ shows that $f(z)$ can be continued analytically in any region in which $\cos \alpha \leq 0$; for $\cos \alpha > 0$ it can be continued up to $\cos \alpha = 1$. In other words analytic continuation is available for $\mathcal{R}(z) < 1$. Actually the continuation has a single pole at $z = 1$ since it is $1/(1 - z)$.

In contrast knowledge of the singularities of $f(z)$ and its analytic continuation gives some information about the growth of φ. For, if $r_0 e^{i\alpha}$ is a singularity of $f(z)$, Eq.(3.2.8) must fail otherwise Eq.(3.2.7) would be convergent. Again, if there are no singularities in $\mathcal{R}(z) < 1$,

$$|\varphi(zt)| < K \exp(tr \cos \alpha)$$

must be valid to guarantee the right convergence of Eq.(3.2.7).

It remains to discuss a_m in the light of Eq.(3.2.4). For future purposes it is convenient to deal with a slightly more general condition.

Theorem 3.2.2 *Let m be unlimited. If*

$$\frac{a_{m+1}}{a_m} = 1 + \frac{a + \varkappa}{m^q}$$

then, if $0 \le q < 1$ with $a = 0$ if $q = 0$,

$$a_m = \exp\{(a + \varkappa)m^{1-q}/(1-q)\}$$

and, if $q = 1$,

$$a_m = \exp\{(a + \varkappa)\ln m\}.$$

Proof. Let m be a positive integer and n an unlimited integer. Then

$$\ln \frac{a_{n+m}}{a_n} = \sum_{p=n}^{n+m-1} \ln \frac{a_{p+1}}{a_p} = \sum_{p=n}^{n+m-1} \ln\left(1 + \frac{a + \varkappa}{p^q}\right).$$

Since p is unlimited $(a + \varkappa)/p^q$ is infinitesimal $(a = 0$ if $q = 0)$. Hence the right-hand side is

$$\sum_{p=n}^{n+m-1} \frac{a + \varkappa}{p^q} = (a + \varkappa) \sum_{p=n}^{n+m-1} \frac{1}{p^q}$$

by Theorem A.5.10. From Theorem A.5.11

$$\int_n^{n+m} \frac{dt}{t^q} < \sum_{p=n}^{n+m-1} \frac{1}{p^q} < \int_{n-1}^{n+m-1} \frac{dt}{t^q}$$

whence, for $q < 1$,

$$(n+m)^{1-q} - n^{1-q} < \sum_{p=n}^{n+m-1} \frac{1-q}{p^q} < (n+m-1)^{1-q} - (n-1)^{1-q}.$$

The two extremes of the inequality differ by an infinitesimal multiple; hence

$$\sum_{p=n}^{n+m-1} \frac{1}{p^q} = \{(n+m)^{1-q} - n^{1-q}\}(1 + \varkappa)/(1-q)$$

when $q < 1$. Therefore

$$\frac{a_{n+m}}{a_n} = \exp\left[\frac{a + \varkappa}{1-q}\left\{(n+m)^{1-q} - n^{1-q}\right\}\right]. \qquad (3.2.9)$$

Let ν be an unlimited integer. For every standard positive integer n, we have $n/\nu^{1-q} \simeq 0$ and $\ln a_n/\nu^{1-q} \simeq 0$. By Robinson's Lemma (Section A.2) there is an unlimited integer N such that $N/\nu^{1-q} \simeq 0$ and $\ln a_N/\nu^{1-q} \simeq 0$. Therefore

$$\ln a_\nu = \ln(a_\nu/a_N) + \ln a_N = \frac{a + \varkappa}{1-q}(\nu^{1-q} - N^{1-q}) + \ln a_N$$

from Eq.(3.2.9). The properties of N ensure that

$$\ln a_\nu = (a + \varkappa)\nu^{1-q}/(1-q).$$

The statement of the theorem is recovered for $q < 1$.

When $q = 1$ the only changes which are needed in the preceding argument are that

$$\sum_{p=n}^{n+m-1} \frac{1}{p} = (a + \varkappa)\ln\{(n+m)/n\}$$

and that ν^{1-q} should be replaced by $\ln\nu$ for the application of Robinson's Lemma.

The theorem is proved. ■

An inequality which can be useful at times can be deduced from Eq.(3.2.9). For

$$\frac{1}{1-q}\{(n+m)^{1-q} - n^{1-q}\} = \int_n^{n+m} \frac{dt}{t^q} < \frac{1}{n^q}\int_n^{n+m} dt < \frac{m}{n^q}$$

which gives

Corollary 3.2.2 *Under the conditions of Theorem 3.2.2*

$$a_{n+m}/a_n < \exp\{(|a| + \varkappa)m/n^q\}.$$

The implications of Theorem 3.2.2 for the summing of terms in the series for $\varphi(z)$ are examined in the next section.

3.3 Partial sums

We turn to the matter of summing some of the terms in Eq.(3.2.6) when the coefficients are subject to Eq.(3.2.4) so that benefit can be drawn from Theorem 3.2.2 with $q = \frac{1}{2}$.

Theorem 3.3.1 *Let $a_{m+1}/a_m = 1 + (a + \varkappa)/m^{\frac{1}{2}}$ when m is unlimited. Let ω be an unlimited positive integer. Then there is an unlimited Δ such that*

$$\sum_{m=\omega}^{\infty} \frac{a_m}{m!}z^m = \frac{1+\varkappa}{1-z/\omega}\cdot\frac{a_\omega z^\omega}{\omega!}$$

for $|z| < \omega + \Delta\omega^{\frac{1}{2}}$ so long as $|1 - z/\omega| \geq \delta > 0$ with δ standard.
Proof. Write

$$\frac{\omega!}{a_\omega z^\omega}\sum_{m=\omega}^{\infty} \frac{a_m}{m!}z^m = \sum_{m=\omega}^{\infty} \frac{\omega! a_m}{m! a_\omega}z^{m-\omega} = \sum_{n=0}^{\infty} b_n\left(\frac{z}{\omega}\right)^n$$

where

$$b_n = \frac{\omega!\omega^n a_{n+\omega}}{(\omega+n)! a_\omega}.$$

Since ω is unlimited the factorials can be estimated by Stirling's formula (see Eq.(2.1.5) of Chapter 2) and the ratio $a_{n+\omega}/a_\omega$ obtained from Eq.(3.2.9) with $q = \frac{1}{2}$. As a consequence

$$b_n = \exp[2(a + \varkappa)\{(\omega+n)^{\frac{1}{2}} - \omega^{\frac{1}{2}}\} + n - (\omega+n+\tfrac{1}{2})\ln(1+n/\omega) + \varkappa]. \quad (3.3.1)$$

Let $|z| \leq \omega + d\omega^{\frac{1}{2}}$ where d is positive and limited. Since (see Corollary 3.2.2) $2\{(\omega + n)^{\frac{1}{2}} - \omega^{\frac{1}{2}}\} < n/\omega^{\frac{1}{2}}$

$$\left| b_n \left(\frac{z}{\omega}\right)^n \right| \leq \exp\left[(|a| + \varkappa)n/\omega^{\frac{1}{2}} + n + 1 - (\omega + n + \tfrac{1}{2})\ln\left(1 + \frac{n}{\omega}\right) \right.$$
$$\left. +n\ln\left(1 + \frac{d}{\omega^{1/2}}\right) \right].$$

Now $\ln x \leq x - 1$ and so, if d' is any standard number greater than $d + |a| + \varkappa$,

$$\left| b_n \left(\frac{z}{\omega}\right)^n \right| \leq \exp\left[n + 1 - (\omega + n + \tfrac{1}{2})\ln(1 + n/\omega) + d'n/\omega^{\frac{1}{2}} \right]. \qquad (3.3.2)$$

Split the series into two parts. One part consists of the terms in which $n \geq \omega'$ whereas in the other part $n \leq \omega' - 1$ where ω' is the next unlimited integer above $\omega^{5/8}$.

Replace n in the exponent of Eq.(3.3.2) by the variable t. Its derivative is

$$\frac{d'}{\omega^{\frac{1}{2}}} - \ln\left(1 + \frac{t}{\omega}\right) - \frac{1/2}{t + \omega}$$

which falls steadily as t increases from 0. When $t = \omega'$ use the fact that, for $x \geq 0$,

$$\ln(1 + x) \geq x - x^2/2; \qquad (3.3.3)$$

then the derivative must be less than $-\omega'/2\omega$ since $\omega' > \omega^{\frac{1}{2}}$ and d' is limited. Hence

$$\frac{d'}{\omega^{\frac{1}{2}}} - \ln\left(1 + \frac{t}{\omega}\right) - \frac{1/2}{t + \omega} < -\frac{\omega'}{2\omega}$$

for $t \geq \omega'$. Integrate the inequality from ω' to n when $n \geq \omega'$ to obtain

$$n + d'n/\omega^{\frac{1}{2}} - (\omega + n + \tfrac{1}{2})\ln(1 + n/\omega)$$
$$< -\frac{\omega'}{2\omega}(n - \omega') + \omega' + d'\omega'/\omega^{\frac{1}{2}} - (\omega + \omega' + \tfrac{1}{2})\ln(1 + \omega'/\omega)$$
$$< -\omega'n/4\omega$$

on employing Eq.(3.3.3) and noting that any positive terms are dominated by $\omega'n/4\omega$.

Insertion of the inequality into Eq.(3.3.2) leads to

$$\sum_{n=\omega'}^{\infty} \left| b_n \left(\frac{z}{\omega}\right)^n \right| < \frac{\exp(1 - \omega'^2/4\omega)}{1 - \exp(-\omega'/4\omega)} < \omega^{3/8}\exp(-\omega^{\frac{1}{4}}/2) \simeq 0. \qquad (3.3.4)$$

Consider now the terms in which $n < \omega'$. It is straightforward to verify that

$$\left(1 - \frac{z}{\omega}\right) \sum_{n=0}^{\omega'-1} b_n \left(\frac{z}{\omega}\right)^n = b_0 - b_{\omega'-1}\left(\frac{z}{\omega}\right)^{\omega'} - \sum_{n=0}^{\omega'-2}(b_n - b_{n+1})\left(\frac{z}{\omega}\right)^{n+1}.$$

Here $b_0 = 1$ by definition and the next term is infinitesimal by what has just been demonstrated. So

$$\left(1 - \frac{z}{\omega}\right) \sum_{n=0}^{\omega'-1} b_n \left(\frac{z}{\omega}\right)^n = 1 + \varkappa - \sum_{n=0}^{\omega'-2} (b_n - b_{n+1}) \left(\frac{z}{\omega}\right)^{n+1}. \qquad (3.3.5)$$

The plan now is to establish that the series on the right of Eq.(3.3.5) is infinitesimal. To this end observe that

$$\left(1 - \frac{z}{\omega}\right) \sum_{n=0}^{\omega'-2} (b_n - b_{n+1}) \left(\frac{z}{\omega}\right)^{n+1} = (b_0 - b_1)\frac{z}{\omega} - (b_{\omega'-2} - b_{\omega'-1}) \left(\frac{z}{\omega}\right)^{\omega'-1}$$

$$- \sum_{n=0}^{\omega'-3} (b_n - 2b_{n+1} + b_{n+2}) \left(\frac{z}{\omega}\right)^{n+2} \qquad (3.3.6)$$

The second term on the right of Eq.(3.3.6) is infinitesimal by what has been shown already. As regards the first term

$$b_0 - b_1 = \frac{1}{\omega + 1} - \frac{\omega^{\frac{1}{2}}(a + \varkappa)}{\omega + 1} \simeq 0$$

on utilising Eq.(3.2.4). Since $|z/\omega| < 1 + d/\omega^{\frac{1}{2}}$ the first term is infinitesimal also. It remains to consider the last term of Eq.(3.3.6).

Since $n < \omega'$ it may be verified readily, by means of Eq.(3.2.4), that

$$b_n - 2b_{n+1} + b_{n+2} = \varkappa b_n / \omega^{\frac{1}{2}}.$$

Substitute from Eq.(3.3.2) with Eq.(3.3.3) incorporated; then

$$\sum_{n=0}^{\omega'-3} \left| (b_n - 2b_{n+1} + b_{n+2}) \left(\frac{z}{\omega}\right)^{n+2} \right| \leq \varkappa \sum_{n=0}^{\omega'-3} \frac{1}{\omega^{\frac{1}{2}}} \exp\left(\frac{nd'}{\omega^{\frac{1}{2}}} - \frac{n^2}{4\omega}\right) \qquad (3.3.7)$$

for some infinitesimal $\varkappa$. Put $h = 1/\omega^{\frac{1}{2}}$ so that the series becomes

$$h \sum_{n=0}^{\omega'-3} \exp(nd'h - n^2h^2/4).$$

Since h is infinitesimal, comparison with Definition A.5.1 reveals that this series represents an integral which is bounded by

$$\int_0^{\infty} \exp(d't - t^2/4)dt.$$

The value of the integral is limited and so the right-hand side of Eq.(3.3.7) is infinitesimal. Hence Eq.(3.3.6) implies that

$$\left(1 - \frac{z}{\omega}\right) \sum_{n=0}^{\omega'-2} (b_n - b_{n+1}) \left(\frac{z}{\omega}\right)^{n+1} = \varkappa.$$

Accordingly, provided that $|1 - z/\omega| \geq \delta > 0$ with δ standard,

$$\sum_{n=0}^{\omega'-2} (b_n - b_{n+1}) \left(\frac{z}{\omega}\right)^{n+1} = \varkappa$$

and it follows from Eq.(3.3.5) that

$$\sum_{n=0}^{\omega'-1} b_n \left(\frac{z}{\omega}\right)^n = (1 + \varkappa)/(1 - z/\omega).$$

The upper limit can be altered to ∞ on account of Eq.(3.3.4) and $|z|/\omega$ being limited. The validity of the theorem has been affirmed for $|z| \leq \omega + d\omega^{\frac{1}{2}}$ with d limited.

The set of integers $n \in \mathbf{N}$ for which

$$\left| \frac{\omega!(1 - z/\omega)}{a_\omega z^\omega} \sum_{m=\omega}^{\infty} \frac{a_m}{m!} z^m - 1 \right| \leq \frac{1}{n}$$

and $(|z| - \omega)/\omega^{\frac{1}{2}} < n$ includes all the standard integers by what has just been proved. There is no set which consists solely of the standard integers. As a consequence there must be an unlimited integer Δ for which these relations hold with Δ in place of n. Since $1/\Delta$ is infinitesimal there is nothing more to prove. ∎

Theorem 3.3.1 approximates the sum of the higher order terms in the series for the entire function $\varphi(z)$ when z lies inside a certain circle. Broadly speaking, an explanation of how the formula originates is the following. To go from one term to the next when n is large multiply by z/n. Successive terms diminish rapidly when $|z|$ is small and the dominant behaviour is dictated by the first of them, namely $a_\omega z^\omega/\omega!$. As $|z|$ increases successive terms become more important until $|z|$ is about the same as ω when the terms resemble those of a geometric series with ratio z/ω. When $|z|$ grows still larger successive terms increase and it is the last one retained which is dominant. Thus, one can expect a theorem, analogous to Theorem 3.3.1, for the contribution of the lower order terms in $\varphi(z)$ when z is outside a suitable circle; it was proved originally by Izumi (1927) in a classical manner.

Theorem 3.3.2 *If ω is an unlimited positive integer and a_m satisfies the same condition as in Theorem 3.3.1*

$$\sum_{m=0}^{\omega-1} \frac{a_m}{m!} z^m = \frac{1 + \varkappa}{z/\omega - 1} \cdot \frac{a_\omega}{\omega!} z^\omega$$

for $|z| \geq \omega$ provided that $|1 - z/\omega| \geq \delta > 0$ with δ standard.
Proof. Set

$$\frac{(\omega - 1)!}{a_{\omega-1} z^{\omega-1}} \sum_{m=0}^{\omega-1} \frac{a_m}{m!} z^m = \sum_{m=0}^{\omega-1} c_m \left(\frac{z}{\omega}\right)^{m-\omega+1}$$

where
$$c_m = \frac{(\omega-1)!a_m}{m!a_{\omega-1}}\omega^{m-\omega+1}.$$

When m is a standard integer utilise Stirling's formula for $(\omega-1)!$ and Theorem 3.2.2 for $a_{\omega-1}$. The modulus of $(z/\omega)^{m-\omega+1}$ does not exceed 1 since $m < \omega - 1$. Hence

$$\left| c_m \left(\frac{z}{\omega}\right)^{m-\omega+1} \right| \leq \frac{|a_m|}{m!}\exp\{(m+\tfrac{1}{2})\ln\omega - \omega + 2(|a| + и)\omega^{\frac{1}{2}}\}$$
$$\leq |a_m|\exp(-\omega/2)/m! \simeq 0$$

since a_m is limited for standard m. Theorem A.1.3 warrants the assertion that

$$\sum_{m=0}^{m'} c_m(z/\omega)^{m-\omega+1} \simeq 0$$

for every standard integer m'. By Robinson's Lemma there is an unlimited integer M such that $M/\omega \simeq 0$ and

$$\sum_{m=0}^{M} c_m(z/\omega)^{m-\omega+1} \simeq 0. \tag{3.3.8}$$

For $m > M$, a_m can be approximated by Theorem 3.2.2 and $m!$ by Stirling's formula with the net result

$$c_m = \exp\left[(m+\tfrac{1}{2})\ln\frac{\omega}{m} + m - \omega + 2(a+и)\{m^{\frac{1}{2}} - (\omega-1)^{\frac{1}{2}}\} + и\right]. \tag{3.3.9}$$

The precise value of M is unknown; to be on the safe side it will be assumed that $M < \omega''$ where ω'' is the next integer above $\omega^{7/8}$. If the value of M were known and was greater than ω'' it might be possible to dispense with some of the subsequent steps.

The function
$$f(t) = \left(1 + \frac{1}{2t}\right)\ln\frac{\omega}{t} + 1 - \frac{\omega}{t} + 2|a|\{(\omega-1)^{\frac{1}{2}} - t^{\frac{1}{2}}\}/t$$

has a derivative
$$f'(t) = \frac{\omega}{t^{\frac{1}{2}}}\left\{1 - \frac{t}{\omega} - \frac{|a|}{\omega^{\frac{1}{2}}}(1+и)\right\}$$

which is positive for $t \leq \omega - \omega'$ where ω' is the next integer above $\omega^{5/8}$. Hence, for $M \leq t \leq \omega''$,

$$f(t) \leq f(\omega'') < -\omega/2\omega'' \tag{3.3.10}$$

whereas, for $t \leq \omega - \omega'$,

$$f(t) \leq f(\omega - \omega') < -(\omega'/2\omega)^2. \tag{3.3.11}$$

Therefore, from Eq.(3.3.10),

$$\sum_{m=M+1}^{\omega''} \left| c_m (z/\omega)^{m-\omega+1} \right| < \sum_{m=M+1}^{\omega''} \exp(-m\omega/2\omega'')$$

$$< \frac{\exp\{-(M+1)\omega/2\omega''\} - \exp(-\omega)}{1 - \exp(-\omega/2\omega'')}$$

$$\simeq 0.$$

On the other hand, Eq.(3.3.11) supplies

$$\sum_{m=\omega''+1}^{\omega-\omega'} \left| c_m (z/\omega)^{m-\omega+1} \right| < \sum_{m=\omega''+1}^{\omega-\omega'} \exp\{-m(\omega'/\omega)^2\}$$

$$< \frac{\exp\{-\omega''(\omega'/2\omega)^2\} - \exp\{-(\omega-\omega')(\omega'/2\omega)^2\}}{1 - \exp\{-(\omega'/2\omega)^2\}}$$

$$\simeq 0$$

since $\omega''(\omega'/\omega)^2 > \omega^{1/8}$ and $\omega(\omega'/\omega)^2 > \omega^{1/4}$.

Consequently, the only part of the series which may not be infinitesimal is

$$\sum_{m=\omega-\omega'+1}^{\omega-1} c_m (z/\omega)^{m-\omega+1} = \sum_{n=0}^{\omega'-2} c_{\omega-1-n}(\omega/z)^n. \tag{3.3.12}$$

Now

$$(1 - \omega/z) \sum_{n=0}^{\omega'-2} c_{\omega-1-n}(\omega/z)^n = 1 + \varkappa + \sum_{n=1}^{\omega'-2} (c_{\omega-1-n} - c_{\omega-n})(\omega/z)^n \tag{3.3.13}$$

and

$$(1 - \omega/z)\left(\sum_{n=1}^{\omega'-2} (c_{\omega-1-n} - c_{\omega-n})(\omega/z)^n \right.$$

$$= (c_{\omega-2} - c_{\omega-1})\omega/z + \varkappa + \sum_{n=2}^{\omega'-2} (c_{\omega-1-n} - 2c_{\omega-n} + c_{\omega-n+1})(\omega/z)^n.$$

$$\tag{3.3.14}$$

The first term on the right of Eq.(3.3.14) is infinitesimal and

$$c_{\omega-1-n} - 2c_{\omega-n} + c_{\omega-n+1} = \varkappa c_{\omega-1-n}/\omega^{\frac{1}{2}}$$

because $n/\omega < \omega^{-3/8}$. Also, from Eq.(3.3.9),

$$\frac{1}{\omega^{\frac{1}{2}}} \sum_{n=2}^{\omega'-2} \left| c_{\omega-1-n} \left(\frac{\omega}{z}\right)^n \right| \le \frac{1}{\omega^{\frac{1}{2}}} \sum_{n=2}^{\omega'-2} \exp\left(4|a| \frac{n}{\omega^{\frac{1}{2}}} - \frac{n^2}{4\omega} \right).$$

As in Theorem 3.3.1 the right-hand side is bounded by a convergent integral and so the right-hand side of Eq.(3.3.14) is infinitesimal.

Hence, so long as $|1 - z/\omega| \geq \delta > 0$,

$$\sum_{n=1}^{\omega'-2} (c_{\omega-1-n} - c_{\omega-n})(\omega/z)^n = \varkappa$$

whence, from Eq.(3.3.12)and Eq.(3.3.13),

$$\sum_{m=\omega-\omega'+1}^{\omega-1} c_m(z/\omega)^{m-\omega+1} = (1+\varkappa)/(1-\omega/z).$$

Accordingly, collecting together all that has been proved, we have

$$\sum_{m=0}^{\omega-1} c_m(z/\omega)^{m-\omega+1} = (1+\varkappa)/(1-\omega/z).$$

The statement of the theorem follows now after invocation of Eq.(3.2.4). ∎

3.4   Asymptotic behaviour of an entire function

The region where $z = \omega$ has been excluded from both Theorems 3.3.1 and 3.3.2. Indeed, the two theorems create the illusion that there is a pole at $z = \omega$. That no singularity can be present is evident from $\varphi(z)$ being an entire function. If both theorems were valid for z close to ω their combination would supply a formula for $\varphi(z)$, namely

$$\varphi(z) = \varkappa a_\omega z^\omega / \{\omega!(1 - z/\omega)\}$$

after two large quantities cancel one another. When $z/\omega \simeq 1$ this suggests that

$$\varphi(\omega) \sim a_\omega e^\omega. \tag{3.4.1}$$

Of course, this discussion does not constitute a proof of Eq.(3.4.1) but, because Eq.(3.4.1) holds when $a_n = 1$ and $\varphi(z) = e^z$, indicates that it could be worth seeing whether or not a proof could be devised. After all, the example could be exceptional and engender a spurious feeling of confidence. The relevant theorem will be demonstrated now; it differs slightly from Eq.(3.4.1).

Theorem 3.4.1 *Under the same condition on a_m as in Theorem 3.3.1*

$$\varphi(z) = (1+\varkappa)a_\omega \exp(z + a^2/2)$$

for $z = \omega + d$ with ω an unlimited positive integer and d limited.

Proof. In the proof of Theorem 3.3.1 the condition $|1 - z/\omega| \geq \delta$ has not been used up to Eq.(3.3.4) and, in fact, is not employed until well after Eq.(3.3.7). Therefore, Eq.(3.3.4) continues to hold and

$$\sum_{n=0}^{\infty} b_n \left(\frac{z}{\omega}\right)^n \simeq \sum_{n=0}^{\omega'-1} b_n \left(\frac{z}{\omega}\right)^n. \qquad (3.4.2)$$

Taking advantage of the smallness of n/ω, we can expand terms in Eq.(3.3.1) to obtain

$$\sum_{n=0}^{\omega'-1} b_n \left(\frac{z}{\omega}\right)^n = \sum_{n=0}^{\omega'-1} \exp\left\{(a + \varkappa)\frac{n}{\omega^{1/2}} - \frac{n^2}{2\omega} + \varkappa\right\}$$

since $z/\omega = 1 + d/\omega$ and d is limited. The sum is estimated by an integral as in Theorem 3.3.1 and leads to consideration of

$$\omega^{\frac{1}{2}} \int_0^{\omega'/\omega^{1/2}} \exp\{(a + \varkappa)t - \tfrac{1}{2}t^2 + \varkappa\}dt.$$

The integrand differs infinitesimally from $\exp(at - \tfrac{1}{2}t^2)$ for limited t. It is bounded by $\exp\{(a + 1)t - \tfrac{1}{2}t^2 + 1\}$ for any t. Hence, by Theorem A.5.5,

$$\sum_{n=0}^{\omega'-1} b_n \left(\frac{z}{\omega}\right)^n = \omega^{\frac{1}{2}}(1 + \varkappa) \int_0^{\infty} \exp(at - \tfrac{1}{2}t^2)dt \qquad (3.4.3)$$

and the upper limit of summation can be replaced by ∞ on account of Eq.(3.4.2).

For the series involving c_n in Theorem 3.3.2 note that Eq.(3.3.13) is reached without invoking $|1 - z/\omega| \geq \delta$. Treat the series in Eq.(3.3.13) in the same way as that with the b_n but starting from Eq.(3.3.9). There results

$$\sum_{m=0}^{\omega-1} c_m(z/\omega)^{m-\omega+1} = \omega^{\frac{1}{2}}(1 + \varkappa) \int_0^{\infty} \exp(-at - \tfrac{1}{2}t^2)dt. \qquad (3.4.4)$$

Combining Eq.(3.4.3) and Eq.(3.4.4) we have

$$\begin{aligned}
\varphi(z) &= \frac{a_\omega}{\omega!} z^\omega \omega^{\frac{1}{2}}(1 + \varkappa) \int_0^{\infty} \exp(at - \tfrac{1}{2}t^2)dt \\
&\quad + \frac{a_{\omega-1}}{(\omega - 1)!} z^{\omega-1} \omega^{\frac{1}{2}}(1 + \varkappa) \int_0^{\infty} \exp(-at - \tfrac{1}{2}t^2)dt \\
&= \frac{a_\omega}{\omega!} z^\omega \omega^{\frac{1}{2}}(1 + \varkappa) \int_{-\infty}^{\infty} \exp(at - \tfrac{1}{2}t^2)dt
\end{aligned}$$

since $\omega a_{\omega-1} = (1 + \varkappa)a_\omega z$. The evaluation of the integral is immediate and

$$\varphi(z) = (1 + \varkappa)(2\pi)^{\frac{1}{2}} a_\omega \omega^{\frac{1}{2}} z^\omega \exp(a^2/2)/\omega!.$$

On substituting Stirling's formula for the factorial the requisite expression is derived and the theorem is proved. ∎

It is convenient to abuse notation a little and denote by a_z the quantity obtained by replacing m by z in the definition of a_m. When $z = \omega + d$ then $a_z = (1 + \varkappa)a_\omega$ by virtue of Theorem 3.2.2. With this convention Theorem 3.4.1 can be stated as

$$\varphi(z) = (1 + \varkappa)a_z \exp(z + \tfrac{1}{2}a^2) \qquad (3.4.5)$$

for unlimited z in the neighbourhood of the real axis. Thus, Eq.(3.4.5) furnishes the asymptotic performance for such z for an entire function whose coefficients have the appropriate behaviour.

When $\varphi(z) = e^z$, $a_n = 1$ and $a = 0$; there is no disagreement with Eq.(3.4.5). As another check on Eq.(3.4.5), suppose that $\varphi(z)$ is the confluent hypergeometric function $M(b, c, z)$ defined by

$$M(b, c, z) = 1 + \frac{b}{c}\frac{z}{1!} + \frac{b(b+1)}{c(c+1)}\frac{z^2}{2!} + \cdots.$$

Here

$$a_m = (b + m - 1)!(c - 1)!/\{(b - 1)!(c + m - 1)!\}$$

so that $a_{m+1}/a_m = 1 + (1 + \varkappa)(b - c)/m$ when m is unlimited but b and c are limited. Consequently, the conditions of Theorem 3.4.1 are met with $a = 0$. Taking advantage of Stirling's formula we have, from Eq.(3.4.5),

$$M(b, c, z) = (c - 1)!z^{b-c}e^z(1 + \varkappa)/(b - 1)!$$

for unlimited z close to the real axis when b and c are limited. This agrees with the usual asymptotic behaviour quoted for the confluent hypergeometric function.

Since Eq.(3.4.5) reproduces the correct formula for the confluent hypergeometric function it will do the same for any function defined in terms of M. For instance, for the modified Bessel function $I_\nu(x)$ it gives

$$\begin{aligned} I_\nu(x) &= \left(\tfrac{1}{2}x\right)^\nu e^{-x} M(\tfrac{1}{2} + \nu, 1 + 2\nu, 2x)/\nu! \\ &= (1 + \varkappa)e^x/(2\pi x)^{\frac{1}{2}} \end{aligned}$$

for unlimited positive x and limited ν. Likewise, we have

$$\begin{aligned} \gamma(b, x) &= \int_0^x t^{b-1}e^{-t}dt = x^b e^{-x} M(1, 1 + b, x)/b \\ &= (b - 1)!(1 + \varkappa) \end{aligned}$$

for unlimited positive x and limited b. Both these asymptotic formulae are in conformity with standard results, though that for $\gamma(b, x)$ is an immediate consequence of the integral representation without any need for the intervention of M.

The sums of other series can be deduced from the foregoing. From Eq.(3.4.5)

$$\sum_{m=0}^{\infty} a_m \frac{\omega^m}{m!} = (1 + и)a_\omega \exp(\omega + a^2/2).$$

Both Theorems 3.3.1 and 3.3.2 are valid when $z = -\omega$. Hence

$$\sum_{m=0}^{\infty} a_m \frac{(-\omega)^m}{m!} = и\, a_\omega e^\omega / \omega^{\frac{1}{2}}.$$

Addition and subtraction of these two equations supplies

$$\sum_{m=0}^{\infty} a_{2m}\omega^{2m}/(2m)! = \tfrac{1}{2}(1+и)a_\omega \exp(\omega + a^2/2),$$

$$\sum_{m=0}^{\infty} a_{2m+1}\omega^{2m+1}/(2m+1)! = \tfrac{1}{2}(1+и)a_\omega \exp(\omega + a^2/2).$$

These formulae can be rewritten in a somewhat more convenient form by adopting the convention of Eq.(3.4.5).

Theorem 3.4.2 *If*

$$\frac{A_{m+1}}{A_m} = 1 + \frac{a + и}{m^{\frac{1}{2}}}$$

for unlimited m then

$$\sum_{m=0}^{\infty} A_m x^{2m}/(2m)! = \tfrac{1}{2}(1+и)A_{\frac{1}{2}x}\exp(x + a^2/4),$$

$$\sum_{m=0}^{\infty} A_m x^{2m+1}/(2m+1)! = \tfrac{1}{2}(1+и)A_{\frac{1}{2}(x-1)}\exp(x + a^2/4)$$

for unlimited positive x.

As an application consider again the modified Bessel function $I_\nu(x)$. It has the series representation

$$I_\nu(x) = \sum_{m=0}^{\infty} \left(\tfrac{1}{2}x\right)^{\nu+2m} /\{m!(\nu + m)!\}.$$

After extraction of the factor $(\tfrac{1}{2}x)^\nu$ the series is in the form of Theorem 3.4.2 with

$$A_m = \frac{(2m)!}{m!(\nu + m)!2^{2m}}.$$

Clearly A_m satisfies the condition of Theorem 3.4.2 and, indeed, it can be checked that $A_m = (1 + и)/\pi^{\frac{1}{2}}m^{\nu+\frac{1}{2}}$ when m is unlimited while ν is limited. Hence

$$I_\nu(x) = (\tfrac{1}{2}x)^\nu(1+и)e^x/2\pi^{\frac{1}{2}}(\tfrac{1}{2}x)^{\nu+\frac{1}{2}} = (1+и)e^x/(2\pi x)^{\frac{1}{2}}$$

in harmony with what has been derived earlier.

Example 3.4.1 *The Airy function*

The Airy functions have the representations

$$\text{Ai}(z) = f(z) - g(z),$$
$$\text{Bi}(z) = \{f(z) + g(z)\}3^{1/2}$$

where

$$f(z) = \frac{1}{(-\frac{1}{3})!3^{\frac{2}{3}}}\left(1 + \frac{z^3}{3!} + \frac{1.4}{6!}z^6 + \frac{1.4.7}{9!}z^9 + \cdots\right),$$

$$g(z) = \frac{1}{(-\frac{2}{3})!3^{\frac{1}{3}}}\left(z + \frac{2}{4!}z^4 + \frac{2.5}{7!}z^7 + \frac{2.5.8}{10!}z^{10} + \cdots\right)..$$

At first sight these series do not appear to conform to our earlier pattern. However, after the substitution $x = (3w/2)^{2/3}$, $f(x)$ becomes a series of even powers of w in which

$$A_m = \frac{(m - \frac{2}{3})!(2m)!3^{3m-\frac{1}{6}}}{(3m)!\pi 2^{2m+1}}$$

on observing that $(-\frac{1}{3})!(-\frac{2}{3})! = 2\pi/3^{\frac{1}{2}}$. For unlimited m

$$A_m = (1 + и)/3^{2/3}\pi^{1/2}m^{1/6}$$

so that Theorem 3.4.2 can be applied with $a = 0$. Hence

$$f(x) = \frac{(1 + и)e^w}{2.3^{2/3}\pi^{\frac{1}{2}}(\frac{1}{2}w)^{1/6}} = \frac{(1 + и)\exp(2x^{\frac{3}{2}}/3)}{2(3\pi)^{\frac{1}{2}}x^{\frac{1}{4}}} \tag{3.4.6}$$

for unlimited positive x. A similar procedure may be followed for $g(x)/x$ and it is found that $g(x)$ is given by Eq.(3.4.6) also.

These results are not much help in determining the asymptotic behaviour of $\text{Ai}(x)$ because the dominant terms cancel. This is not too surprising since it has been shown already in Section 3.1 that $\text{Ai}(x)$ is exponentially damped (see Eq.(3.1.7)). In contrast, Eq.(3.4.6) is useful for $\text{Bi}(x)$ and

$$\text{Bi}(x) = \frac{1 + и}{\pi^{\frac{1}{2}}x^{\frac{1}{4}}}\exp(2x^{\frac{3}{2}}/3)$$

for unlimited positive x.

It must be borne in mind that the formulae which have been established are based on ω being unlimited. When it comes to numerical values with ω in the region of 10 it is not so clear that a theoretically infinitesimal quantity will be truly negligible. We can expect the theorems of this section to be tolerably accurate but it is not so obvious that we can have the same confidence in Theorems 3.3.1 and 3.3.2 as z approaches ω because of the consequent small denominator. Therefore, it is of interest to investigate how Theorems 3.3.1 and 3.3.2 fare numerically as the phase and magnitude of z vary. Multiply the equations in those theorems by factors so that the right-hand side becomes $1 + \varkappa$. Then the left-hand side should be about 1 for all z and ω satisfying the conditions of the theorems. As a test values of the ratio for $\varphi(z) = e^z$ were calculated for $\omega = 10$ and $\omega = 20$. These values are pretty close to the margin when regarded as unlimited numbers and form a severe trial for the theorems. Table 3.4.1 gives some results for positive real z. As can be seen Theorem 3.3.1 is better than 10% accurate when z is less than $\omega/2$ but the performance deteriorates rapidly for larger z. In contrast, the accuracy of Theorem 3.3.2 improves as z increases, being better than 10% when z exceeds 2ω. To put it another way, for these relatively low values of ω, z must be a reasonable distance away from ω to secure accuracy.

Table 3.4.1 Ratio for real z

	Th. 3.3.1			Th. 3.3.2	
z	$\omega = 10$	$\omega = 20$	z	$\omega = 10$	$\omega = 20$
2	0.97404	0.99219	11	0.28524	
5	0.87764	0.98027	15	0.71850	
9	0.34794	0.94174	21	0.87284	0.22158
15		0.74612	25	0.91022	0.64321
19		0.25379	30	0.93541	0.81550
			40	0.95942	0.91826

The degeneration in performance is caused by z approaching ω with insufficient terms in Theorem 3.3.2 to offset $1 - z/\omega$ being nearly zero. The situation is much better when there is no possibility of z coalescing with ω. In Table 3.4.2 the same ratio is calculated for a purely imaginary z. Now the performance is highly satisfactory even when $|z| = \omega$.

Table 3.4.2 Ratio for imaginary z

Th. 3.3.1

z	$\omega = 10$	$\omega = 20$
$2i$	$1.00580 - 0.01683i$	$1.00073 - 0.00488i$
$5i$	$1.02894 - 0.02743i$	$1.00495 - 0.01026i$
$10i$	$1.05547 - 0.00398i$	$1.01533 - 0.01294i$
$15i$		$1.02350 - 0.00813i$
$20i$		$1.02633 - 0.00078i$

Th. 3.3.2

z	$\omega = 10$	$\omega = 20$
$10i$	$1.05497 - 0.00387i$	
$15i$	$1.04758 + 0.02018i$	
$20i$	$1.03376 + 0.02740i$	$1.02633 - 0.00078i$
$25i$	$1.02400 + 0.02766i$	$1.02521 + 0.00540i$
$30i$	$1.01763 + 0.025941i$	$1.02240 + 0.00942i$

Functions which are not entire are not covered by Theorems 3.3.1 and 3.3.2. Notwithstanding, there is a theorem similar to Theorem 3.3.2.

Theorem 3.4.3 *If $|z| > 1 + \delta$, where δ is positive and standard,*

$$\sum_{m=0}^{\omega-1} a_m z^m = (1 + \varkappa)a_\omega z^\omega /(z - 1)$$

for unlimited ω.

Proof. Write

$$\frac{1}{a_{\omega-1}z^{\omega-1}} \sum_{m=0}^{\omega-1} a_m z^m = \sum_{m=0}^{\omega-1} d_m z^{m-\omega+1}$$

with $d_m = a_m/a_{\omega-1}$. When m is standard $|z|^{m-\omega+1}$ is sufficiently small to make the term infinitesimal. Hence, there is an unlimited integer M possessing the properties $M/\omega \simeq 0$ and

$$\sum_{m=0}^{M} d_m z^{m-\omega+1} \simeq 0.$$

Also

$$\sum_{m=M+1}^{\omega-1} d_m z^{m-\omega+1} = \sum_{n=0}^{\omega-M-2} d_{\omega-1-n} z^{-n}.$$

For limited n, $d_{\omega-n-1} \simeq 1$; in addition, for any n,

$$\left| d_{\omega-1-n} z^{-n} \right| \leq \left(\frac{1 + \varkappa}{1 + \delta} \right)^n.$$

Since $(1 + \varkappa)/(1 + \delta) < 1$ this is a convergent sequence. Hence, by Theorem A.5.9.

$$\sum_{n=0}^{\omega-M-2} d_{\omega-1-n} z^{-n} = \frac{1 + \varkappa}{1 - 1/z} \left(1 - \frac{1}{z^{\omega-M-1}} \right) = \frac{(1 + \varkappa)z}{z - 1}.$$

The proof can be completed now in an obvious manner. ∎

3.5 Partial sums of integrals

The theorems which have just been established permit estimates of the partial sums of

$$f(z) = \sum_{m=0}^{\infty} a_m z^m.$$

The early terms have been dealt with in Theorem 3.4.3 so that the late terms are of concern here.

Theorem 3.5.1 *If $f(z)$ can be continued analytically across the arc of its circle of convergence at $e^{i\alpha}$ (α standard and $0 < \alpha < 2\pi$) there is $M' > 1$ such that, for $|z| \le M'$ and $\mathrm{ph}\, z = \alpha$,*

$$f(z) = \sum_{m=0}^{\omega-1} a_m z^m + (1+\varkappa)\frac{a_\omega}{\omega!}z^\omega J(\omega, -z/\omega)$$

where

$$J(\mu, z) = \int_0^\infty \frac{t^\mu e^{-t}}{1+zt}dt$$

and ω is unlimited.

The theorem can be stated alternatively as

$$\sum_{m=\omega}^\infty a_m z^m = (1+\varkappa)\frac{a_\omega}{\omega!}z^\omega J(\omega, -z/\omega) \qquad (3.5.1)$$

which complements the result in Theorem 3.4.3. The restriction on α is to ensure that J does not have a singularity on the interval of integration. Note, however, that, if the condition in Eq.(3.5.2) below holds for $\alpha = 0$, it is valid for any α because all the terms in $\varphi(z)$ have the same sign when $\alpha = 0$.

Proof. The discussion of Section 3.2 indicates that the assumption on the analytic continuation of f is equivalent to assuming that

$$\left|\varphi(te^{i\alpha})\right| < Ke^{Mt} \qquad (3.5.2)$$

for $t \ge 0$ and $M < 1$. Also

$$\begin{aligned}
f(z) &= \int_0^\infty e^{-t}\varphi(zt)dt \\
&= \sum_{m=0}^{\omega-1} a_m z^m + \int_0^\infty e^{-t} \sum_{m=\omega}^\infty \frac{a_m}{m!}(zt)^m dt. \qquad (3.5.3)
\end{aligned}$$

By virtue of Theorems 3.3.1 and 3.3.2

$$\sum_{m=\omega}^\infty \frac{a_m}{m!}z^m = \begin{cases} \dfrac{1+\varkappa}{1-z/\omega}\dfrac{a_\omega z^\omega}{\omega!} & (|z| < \omega + \Delta\omega^{\frac{1}{2}}) \\[2ex] \dfrac{1+\varkappa}{1-z/\omega}\dfrac{a_\omega z^\omega}{\omega!} + \varphi(z) & (|z| \ge \omega). \end{cases}$$

Hence the integral in Eq.(3.5.3) can be expressed as

$$\frac{a_\omega}{\omega!}z^\omega \int_0^\infty \frac{(1+\varkappa)t^\omega}{1-zt/\omega}e^{-t}dt + \int_{(\omega+\Delta\omega^{1/2})/|z|}^\infty e^{-t}\varphi(zt)dt.$$

On account of Eq.(3.5.2) the second integral converges absolutely if $|z|\,M < 1$. This can be arranged by a suitable choice of M', say $M' = (1 + 1/M)/2$. Since the lower limit of integration is an unlimited positive number the integral is infinitesimal. The first integral becomes that stated in the theorem on application of Theorem A.5.5. The proof is finished. ∎

An integral like J has been examined in Theorem 2.1.1 of Chapter 2. Put there $\epsilon = z/\omega$ and $\mu = \omega$; then ϵ is infinitesimal and $\mu\epsilon$ limited since z is limited. Accordingly

$$J(\omega, -z/\omega) = \omega!(1 + \varkappa)/(1 - z).$$

Substitution in Theorem 3.5.1 gives

Corollary 3.5.1 *Under the conditions of Theorem 3.5.1*

$$f(z) = \sum_{m=0}^{\omega-1} a_m z^m + (1 + \varkappa)a_\omega z^\omega/(1 - z).$$

Eq.(3.5.1) may be modified correspondingly.

Instead of integrating along the real axis we can consider integration along a ray in the complex plane as in Theorem 3.1.2. Apart from changing the phase of z to $\alpha - \beta$ with $|\beta| \leq \frac{1}{2}\pi - \delta$ the only alteration to the formula in Theorem 3.5.1 is that the integral in J is along the ray. The path of integration can be deformed into the positive real axis. A potential pole at $t = 1/z$ may be captured during the process. Indeed, with ph $z = \alpha - \beta$ and $0 < \alpha < 2\pi$ but z not positive real,

$$\int_0^{\infty e^{i\beta}} \frac{t^\mu e^{-t}}{1 - zt}dt = J(\mu, -z) + \frac{2\pi i}{z^{\mu+1}}e^{-1/z}\{H(\beta)H(\beta - \alpha)$$
$$-e^{2\pi\mu i}H(-\beta)H(\alpha - \beta - 2\pi)\} \qquad (3.5.4)$$

where $H(x)$ is the Heaviside step function which is 1 for $x > 0$ and 0 for $x < 0$.

By means of Eq.(3.5.4) we have

Theorem 3.5.2 *Subject to Eq.(3.5.2) there is $M' > 1$ such that, for $|z| \leq M'$ with ph $z = \alpha - \beta$ where $0 < \alpha < 2\pi$ and $|\beta| < \frac{1}{2}\pi - \delta$,*

$$\int_0^{\infty e^{i\beta}} e^{-t}\varphi(zt)dt = \sum_{m=0}^{\omega-1} a_m z^m + (1 + \varkappa)\frac{a_\omega}{\omega!}z^\omega\Big[J(\omega, -z/\omega)$$
$$+2\pi i\left(\frac{\omega}{z}\right)^{\omega+1}e^{-\omega/z}\{H(\beta)H(\beta - \alpha)$$
$$-e^{2\pi\omega i}H(-\beta)H(\alpha - \beta - 2\pi)\}\Big]$$

provided that z is not positive real.

There is a theorem similar to Theorem 3.5.1 for the integral of $f(z)$.

Theorem 3.5.3 *Let $|f(te^{i\alpha})| < Ae^{\gamma t}$ for $t \geq 0$ and some α satisfying the inequality $0 < \alpha < 2\pi$ with A and γ standard. Then, for infinitesimal ϵ with $\text{ph}\,\epsilon = \alpha$,*

$$\int_0^\infty e^{-t}f(\epsilon t)dt = \sum_{m=0}^{n-1} m!a_m\epsilon^m + a_n\epsilon^n(1+\varkappa)J(n,-\epsilon)$$

for every $n \in \mathbf{N}$.

Proof. Suppose, firstly, that n is unlimited. The assumption on f implies that f can be continued analytically across the arc of the circle of convergence at $e^{i\alpha}$. Therefore, by Corollary 3.5.1 and Theorem 3.3.3,

$$\sum_{m=n}^\infty a_m z^m = \begin{cases} \dfrac{(1+\varkappa)a_n z^n}{1-z} & (|z| \leq M') \\ \dfrac{(1+\varkappa)a_n z^n}{1-z} + f(z) & (|z| > 1+\delta). \end{cases}$$

Choose $\delta < M'$ and then

$$\int_0^\infty e^{-t}\sum_{m=n}^\infty a_m(\epsilon t)^m dt = a_n\epsilon^n\int_0^\infty \frac{(1+\varkappa)t^n}{1-\epsilon t}e^{-t}dt + \int_{M'/|\epsilon|}^\infty e^{-t}f(\epsilon t)dt.$$

The second integral on the right is absolutely convergent and the lower limit is unlimited; it is infinitesimal therefore. The first integral can be written as in the theorem which is verified now for unlimited n.

When n is limited Theorem 3.1.1 shows that the remainder after the first n terms is $(n!a_n + \varkappa)\epsilon^n$. But, by Theorem 2.1.1 of Chapter 2,

$$J(n,-\epsilon) = n!(1+\varkappa)/(1-n\epsilon) = n!(1+\varkappa)$$

since n is limited and ϵ is infinitesimal. Thus, agreement with the theorem has been secured and there is nothing more to prove. ∎

Example 3.5.1 Let

$$f(z) = (1-z)^s$$

and suppose that s is not a positive integer. Then

$$a_n = s!(-)^n/n!(s-n)!$$

so that $a_{n+1}/a_n \sim 1 - (s+1)/n$. The conditions that have been assumed are satisfied. Therefore, so long as ϵ is not positive real,

$$\int_0^\infty e^{-t}t^\mu(1-\epsilon t)^s dt = \sum_{m=0}^{n-1}\frac{(m+\mu)!s!}{m!(s-m)!}(-\epsilon)^m + \frac{s!(-\epsilon)^n}{n!(s-n)!}(1+\varkappa)J(\mu+n,-\epsilon).$$

Remark that if s is a negative integer the quantity $s!/(s-m)!$ should be replaced by $(m-1-s)!(-)^m/(-s-1)!$.

If the same substitution for J as in Corollary 3.5.1 is made and the infinitesimal $\varkappa$ dropped the remainder can be expressed in the form

$$R_n(\epsilon) = \frac{s!(n+\mu)!(-\epsilon)^n}{n!(s-n)!\{1+(n+\mu)\epsilon\}}.$$

It is of interest to examine how the series and remainder behave as n varies with ϵ fixed. The first illustration is

$$\left(-\frac{\epsilon}{\pi}\right)^{\frac{1}{2}} \int_0^\infty e^{-t} t^{-\frac{1}{2}} (1-\epsilon t)^{-\frac{1}{2}} dt$$

which is related to a modified Bessel function. For $\epsilon = -1/10$ the sum of the series has been calculated for various values of n and displayed in Table 3.5.1. Only sufficient figures are kept in the sum to reach the digit where the remainder would have an effect. Also shown are the corresponding R_n and what the remainder should be to achieve the correct answer of 0.30906732 as well as their ratio. Both R_n and the exact remainder exhibit the same kind of behaviour. They reduce steadily in magnitude as n increases until n reaches 11; thereafter their magnitudes increase with n. The divergence of the asymptotic series is clearly evident at $n = 35$. A deduction from the table is that the asymptotic series gives its best approximation when $|n\epsilon|$ is about 1. It is transparent also that the inclusion of R_n is worthwhile. It always improves the accuracy of the approximation by one figure and sometimes by two or three since the error in R_n never exceeds $\frac{1}{2}\%$ between $n = 9$ and $n = 20$.

Table 3.5.1 Comparison of remainders for $\epsilon = -1/10$

n	Series	R_n	Exact	Ratio
1	0.316228	-0.00753	-0.00716	1.052
5	0.309083	-0.00001585	-0.00001560	1.016
6	0.309060	7.475×10^{-6}	7.386×10^{-6}	1.012
7	0.309072	-4.238×10^{-6}	-4.201×10^{-6}	1.009
8	0.309065	2.810×10^{-6}	2.792×10^{-6}	1.006
9	0.309069	-2.134×10^{-6}	-2.125×10^{-6}	1.004
10	0.309065	1.827×10^{-6}	1.822×10^{-6}	1.003
11	0.309069	-1.742×10^{-6}	-1.740×10^{-6}	1.001
12	0.309065	1.830×10^{-6}	1.831×10^{-6}	0.999
13	0.309069	-2.102×10^{-6}	-2.104×10^{-6}	0.999
14	0.309065	2.620×10^{-6}	2.625×10^{-6}	0.998
15	0.309071	-3.522×10^{-6}	-3.532×10^{-6}	0.997
20	0.309026	$+0.0000410$	$+0.0000412$	0.995
35	67.0091	-66.306	-66.7	0.994

To see if these features are reproduced for other μ and s we consider now the confluent hypergeometric function

$$\{1/(-\tfrac{1}{6})!\} \int_0^\infty e^{-t} t^{-\frac{1}{6}} (1 - \epsilon t)^{-\frac{1}{6}} dt$$

for $\epsilon = -1/16$. The correct answer is 0.9918367992 and the deviations of the sum of the series are shown in Table 3.5.2. The ratio of the remainders is not given since it scarcely varies from 0.990 throughout the table except for $n = 4$ where it is 1.000. Again R_n and the exact remainder have minima in their magnitudes for $|n\epsilon|$ near 1. Also R_n offers improvement to the estimate provided by the asymptotic series but not as much as in Table 3.5.1 because R_n and the exact remainders are not as close as in Table 3.5.1.

Table 3.5.2 Comparison of remainders for $\epsilon = -1/16$

n	Series	R_n	Exact
4	0.99182545	$+0.000011354$	$+0.000011349$
11	0.991836819	-1.933×10^{-8}	-1.948×10^{-8}
12	0.991836786	1.282×10^{-8}	1.293×10^{-8}
13	0.991836809	-9.294×10^{-9}	-9.376×10^{-9}
14	0.9918367918	7.303×10^{-9}	7.371×10^{-9}
15	0.9918368054	-6187×10^{-9}	-6.247×10^{-9}
16	0.9918367935	5.622×10^{-9}	5.677×10^{-9}
17	0.9918368047	5.453×10^{-9}	-5.508×10^{-9}
18	0.9918367935	5.625×10^{-9}	5.683×10^{-9}
19	0.9918368054	-6.149×10^{-9}	-6.213×10^{-9}
20	0.9918367920	7.101×10^{-9}	7.175×10^{-9}
21	0.9918368079	-8.638×10^{-9}	-8.729×10^{-9}
25	0.9918368299	-3.035×10^{-8}	-3.067×10^{-8}
40	0.991196	$+0.000634$	$+0.000640$

In both cases the sum of the asymptotic series oscillates about the correct value. However, this is a special result due to the terms being real and successively alternating in sign. It cannot be expected to hold in general whereas the qualitative behaviour of the remainders may be of wider applicability. This point will be returned to in Section 3.7.

3.6   Maclaurin series

Example 3.5.1 could be regarded as a particular example of integrating a Maclaurin series. In this section we shall examine what can be said about the remainder in the integral of the Maclaurin series of a function under certain hypotheses.

For instance, it will be assumed throughout that f possesses as many derivatives as are required to justify any expansion.

Theorem 3.6.1 *Let the derivatives of f be such that, for unlimited n,*

$$\left| f^{(n)}(te^{i\alpha}) \right| \le \frac{(b+n)! K}{(p+q\,|t|)^{n+1}}$$

where K, b, p, q are standard constants independent of n with p and q positive. Then, if $\mathrm{ph}\,\epsilon = \alpha$ and ϵ is infinitesimal,

$$\int_0^\infty e^{-t} t^\mu f(\epsilon t) dt = \sum_{m=0}^{n-1} \frac{(m+\mu)!}{m!} f^{(m)}(0) \epsilon^m + O\left\{ \left(\frac{p}{|\epsilon|} \right)^{b+\mu+\frac{1}{2}} \exp\left(-\frac{p}{|\epsilon|} \right) \right\}$$

for n an unlimited integer near to $p/|\epsilon|$ i.e. $n = p(1+\varkappa)/|\epsilon|$. There are no smaller remainders.

Proof. It is well known that

$$f(x) = \sum_{m=0}^{n-1} \frac{x^m}{m!} f^{(m)}(0) + x^n s_n(x)$$

where

$$s_n(x) = \frac{1}{(n-1)!} \int_0^1 (1-u)^{n-1} f^{(n)}(ux) du.$$

Hence

$$\int_0^\infty e^{-t} t^\mu f(\epsilon t) dt = \sum_{m=0}^{n-1} \frac{(m+\mu)!}{m!} f^{(m)}(0) \epsilon^m + \epsilon^n \int_0^\infty e^{-t} t^{\mu+n} s_n(\epsilon t) dt.$$

Now, if we insert the hypothesised bound on $f^{(n)}$ in s_n, we have

$$|s_n(x)| \le \frac{(b+n)! K}{(n-1)!} \int_0^1 \frac{(1-u)^{n-1}}{(p+q\,|x|\,u)^{n+1}} du.$$

The substitution $(p+q)u = (p+qu)v$ supplies

$$\int_0^1 \frac{u^\mu (1-u)^{\nu-1}}{(p+qu)^{\mu+\nu+1}} du = \frac{\mu! (\nu-1)!}{(\mu+\nu)! p^\nu (p+q)^{\mu+1}}. \tag{3.6.1}$$

Accordingly

$$|s_n(x)| \le \frac{(b+n)! K}{n! p^n (p+q\,|x|)}.$$

Therefore

$$\begin{aligned}
\left| \epsilon^n \int_0^\infty e^{-t} t^{\mu+n} s_n(\epsilon t) dt \right| &\le \frac{(b+n)! K}{n! p^n} |\epsilon|^n \int_0^\infty \frac{e^{-t} t^{\mu+n}}{p+q\,|\epsilon|\,t} dt \\
&\le \frac{(b+n)! K}{n! p^{n+1}} (\mu+n)! \, |\epsilon|^n.
\end{aligned}$$

The performance of this bound for unlimited n can be inferred by implementing Stirling's formula. This results in an exponential of

$$(\mu + b + n + \tfrac{1}{2}) \ln n - n + n \ln(|\epsilon|/p).$$

The derivative with respect to n of this expression is

$$\ln(n|\epsilon|/p) + (\mu + p + \tfrac{1}{2})/n$$

which vanishes when

$$\ln(n|\epsilon|/p) = -(\mu + p + \tfrac{1}{2})/n \simeq 0.$$

Hence $n|\epsilon|/p \simeq 1$ and it is evident that this value furnishes a minimum of the exponent.

If we put $n = p/|\epsilon|$ in the exponent the remainder is bounded by

$$\frac{K}{p} \left(\frac{p}{|\epsilon|} \right)^{\mu + b + \frac{1}{2}} \exp\left(-\frac{p}{|\epsilon|} \right).$$

Strictly, taking $n = p/|\epsilon|$ may not be possible because n must be an integer. This can be remedied by putting $n = p/|\epsilon| + \theta$ where $|\theta| < 1$ if the integer nearest to $p/|\epsilon|$ is chosen. The only effect is to multiply the preceding expression by e^θ which does not alter the exponential decay. The form of the remainder stated in the theorem has been obtained and there is nothing more to prove. ∎

The conditions on f can be relaxed. For example, if $f(t)$ is replaced by $tf(t)$ the only change is to multiply by ϵ and increase μ by 1. The net result is to leave the ϵ-dependence of the bound for the remainder unchanged. We state this more generally in the following corollary.

Corollary 3.6.1 *If $g(t) = P_m(t)f(t)$ where $P_m(t)$ is a polynomial of degree m (a standard integer) and f has the same properties as in Theorem 3.6.1 the order of the remainder for*

$$\int_0^\infty e^{-t} t^\mu g(\epsilon t) dt$$

is the same as in Theorem 3.6.1.

Another version can be obtained by replacing $f(t)$ by $e^{ct}f(t)$ where c is standard. Repetition of the analysis of Theorem 3.6.1 with $\exp\{(c\epsilon - 1)t\}$ in place of e^{-t} leads to $\epsilon/(1 - c\epsilon)$ being substituted for ϵ and everything multiplied by $1/(1 - c\epsilon)^\mu$.

Corollary 3.6.1a *With f satisfying the same conditions as in Theorem 3.6.1*

$$(1 - c\epsilon)^\mu \int_0^\infty e^{-t} t^\mu e^{c\epsilon t} f(\epsilon t) dt$$

has the same expansion as in Theorem 3.6.1 but with $\epsilon/(1 - c\epsilon)$ in place of ϵ.

If $f(x) = (1 + x)^s$ it is transparent that the conditions of Theorem 3.6.1 are met for $s \leq -1$ with $p = q = 1$. For larger s advantage can be taken of Corollary 3.6.1. Therefore, the remainders in the integrals should be exponentially small when n is near $1/|\epsilon|$ i.e. $n = 10$ in Table 3.5.1 and $n = 16$ in Table 3.5.2. That this is the case can be checked easily.

3.7   Optimal remainders

It has been seen that the least remainder in the integral of the Maclaurin series occurs when the number of terms in the asymptotic expansion is about $1/|\epsilon|$. This is in distinct contrast to what happens to the entire function $\varphi(z)$. Its series is convergent everywhere and so the remainder can be made smaller always by including enough extra terms in the truncated series. The same is true of the series for $f(z)$ so long as z is in the circle of convergence. The reason for the difference is the (in general) divergent nature of the asymptotic series. In a divergent series the remainder cannot become arbitrarily small as the number of terms in the truncated series grows otherwise the series would be convergent. Inasmuch as earlier examples have exhibited a minimum in the remainder when the series diverges it is worth examining whether there are more general occurrences of the phenomenon.

A typical instance is the remainder in Theorem 3.5.3. Replace J as in Corollary 3.5.1 and the remainder R_n is given by

$$R_n(\epsilon) = n!(1 + \varkappa)a_n \epsilon^n / (1 - n\epsilon) \qquad (3.7.1)$$

so long as $n\epsilon$ is limited and separated from 1 by at least a standard distance. In order to gain an idea of whether the remainder diminishes as the number of terms in the series augments take n to be the unlimited integer ω. The factorial can be estimated by means of Stirling's theorem. As for a_ω, recall that a_n is subject to the conditions of Theorem 3.3.1 and therefore available from Theorem 3.2.2. Perforce

$$|R_n(\epsilon)| \simeq \exp\{(\omega + \tfrac{1}{2})\ln \omega - \omega + \omega \ln |\epsilon| + 2|a + \varkappa|\omega^{\frac{1}{2}}\}/|1 - \omega\epsilon|.$$

The derivative with respect to ω of the exponent is

$$\ln(\omega |\epsilon|) + 1/2\omega + |a + \varkappa|/\omega^{\frac{1}{2}} \simeq \ln(\omega |\epsilon|). \qquad (3.7.2)$$

Thus the derivative is negative when $\omega |\epsilon| < 1$ and positive when $\omega |\epsilon| > 1$. Hence, as $\omega |\epsilon|$ increases and passes through 1, $|R_\omega|$ goes through a minimum and thereafter grows steadily. In consequence the asymptotic series is divergent.

To get the minimum of $|R_\omega|$ we need to make the left-hand side of Eq.(3.7.2) zero. However, that may not be possible to achieve exactly because ω must be an integer. Nevertheless, for any suitable integer ω,

$$\omega \, |\epsilon| = 1 + и. \tag{3.7.3}$$

So the nearest to the minimal remainder is

$$|R_\omega(\epsilon)| = \exp\{-(1 + и)/\, |\epsilon|\}/\, |1 - \omega\epsilon| \tag{3.7.4}$$

where ω satisfies Eq.(3.7.3). It might be said to be an optimal remainder. Note that an optimal remainder is exponentially damped in conformity with Theorem 3.6.1.

Notice that if $a_\omega = \exp(и\omega)$ the argument would be unchanged; for, both Eq.(3.7.3) and Eq.(3.7.4) would be valid still.

Observe also that calculating the minimum of R_ω is actually the same as finding the minimum of $\omega! \, |a_\omega \epsilon^\omega|$. But a typical term in the truncated series is $m!a_m\epsilon^m$ so the first term being rejected by the truncation can be regarded as almost minimal.

These results are brought together in the following theorem.

Theorem 3.7.1 *Let $a_{n+1}/a_n = 1 + и$ for unlimited n. Then the expansion $\sum_{m=1}^{n-1} m!a_m\epsilon^m + R_n(\epsilon)$, where R_n is given by Eq.(3.7.1), has an optimal remainder and almost minimal rejected first term when $n \, |\epsilon| \simeq 1$ and both are exponentially damped as in Eq.(3.7.4).*

What this means in practice is that, in an expansion which satisfies the conditions of Theorem 3.7.1, you should stop adding terms to the series when the smallest is reached. Naturally, if several are of the same size, some criterion will have to be adopted to settle which should be the stopping place. The point may be emphasised by drawing benefit from the properties of the derivative in Eq.(3.7.2). What this tells us is that, if $\omega < 1/\, |\epsilon|$ and not near $1/\, |\epsilon|$, increasing ω by 1 adds a term to the series which is larger in magnitude than its successor and reduces the remainder at the same time. On the other hand, if $\omega > 1/\, |\epsilon|$ and well away from $1/\, |\epsilon|$, an increase of 1 in ω adds a term to the series which is smaller than its successor and simultaneously increases the remainder.

Of course, it must be realised that the rule of stopping at the smallest term refers to those with an index of the order of $1/\, |\epsilon|$. It does not apply to the early terms in the series which can behave in a pretty irregular fashion. Ceasing to add terms to the series during this irregular behaviour could produce errors which would not be corrected adequately by an estimate of the remainder.

Another aspect is to compare as z varies the remainders in the various functions that have been integrated. According to Theorem 3.3.1 the remainder $\rho_n(z)$ in $\varphi(z)$ is given by

$$\rho_\omega(z) = \frac{1 + и}{1 - z/\omega} \frac{a_\omega z^\omega}{\omega!}$$

for $|z| < \omega + \Delta\omega^{\frac{1}{2}}$. The remainder $r_n(z)$ for $f(z)$ is, when $|z| < M'$,

$$r_\omega(z) = \frac{1+\varkappa}{1-z} a_\omega z^\omega$$

from Corollary 3.5.1. Thus, if $|z| < M'$,

$$\frac{\rho_\omega(z)}{r_\omega(z)} = \frac{(1+\varkappa)(1-z)}{\omega!}$$

which shows that the ratio is virtually constant provided that z does not vary too widely and stays away from 1. Also disclosed is that, for given z, many more terms of f than φ are required to attain the same level of remainder consistent with the series for φ being the more rapidly convergent. Likewise

$$\frac{r_\omega(\epsilon)}{R_\omega(\epsilon)} = \frac{(1+\varkappa)(1-\omega\epsilon)}{\omega!}$$

performs in much the same way.

Exercises on Chapter 3

1. If $a_{m+1}/a_m = -1 + (a+\varkappa)/m^{\frac{1}{2}}$ for unlimited m show that

$$\sum_{m=\omega}^{\infty} \frac{a_m}{m!} z^m = \frac{1+\varkappa}{1+z/\omega} \cdot \frac{a_\omega z^\omega}{\omega!}$$

for $|z| < \omega + \Delta\omega^{\frac{1}{2}}$ with $|1 + z/\omega| \geq \delta > 0$.

2. If $a_{m+1}/a_m = c + (a+\varkappa)/m^{\frac{1}{2}}$ for unlimited m with c standard show that

$$\sum_{m=\omega}^{\infty} \frac{a_m}{m!} z^m = \frac{1+\varkappa}{1-cz/\omega} \cdot \frac{a_\omega z^\omega}{\omega!}$$

for $|cz| < \omega + \Delta\omega^{\frac{1}{2}}$ with $|1 - cz/\omega| \geq \delta > 0$. Deduce conditions under which

$$\sum_{m=\omega}^{\infty} \frac{a_m}{m!} z^{m\lambda} = \frac{1+\varkappa}{1-cz^\lambda/\omega} \cdot \frac{a_\omega z^{\omega\lambda}}{\omega!}.$$

3. By changing z to $-z$ in Theorem 3.3.1 and adding the two series show that, for $|z| < 2\omega + \Delta\omega^{\frac{1}{2}}$,

$$\sum_{m=\omega}^{\infty} \frac{a_{2m}}{(2m)!} z^{2m} = \frac{1+\varkappa}{1-(z/2\omega)^2} \cdot \frac{a_{2\omega} z^{2\omega}}{(2\omega)!}.$$

Try replacing z by $\zeta z, \zeta^2 z, \ldots$ where $\zeta = \exp(2\pi i/p)$, p being a positive integer, to see if you can prove that

$$\sum_{m=\omega}^{\infty} \frac{a_{pm} z^{pm}}{(pm)!} = \frac{1 + \varkappa}{1 - (z/p\omega)^p} \cdot \frac{a_{p\omega} z^{p\omega}}{(p\omega)!}.$$

4. Is it possible to change the condition in Theorem 3.3.2 to $|z| \geq \omega - d$, where d is limited and positive, instead of $|z| \geq \omega$ without destroying its validity?

5. If $|f(te^{i\alpha})| \leq Ae^{\gamma t}$ for $t \geq 0$ and ϵ is infinitesimal with $\mathrm{ph}\,\epsilon = \alpha/\lambda$ show that

$$\int_0^{\infty} e^{-t} f(\epsilon^\lambda t) dt = \sum_{m=0}^{n-1} m! a_m \epsilon^{m\lambda} + a_n \epsilon^{n\lambda} (1 + \varkappa) J(n, -\epsilon^\lambda)$$

for $n \in \mathbf{N}$, stating any conditions imposed on α.

6. If $b \geq 0$ prove that

$$\int_0^{\infty} \frac{t^\mu e^{-zt}}{1 + bt} dt = \frac{1}{z^{\mu+1}} J(\mu, b/z)$$

when z is a positive real number. Deduce by analytic continuation that the relation holds for $|\mathrm{ph}\,z| \leq \pi/2 - \delta$ with δ standard and positive.

If the upper limit is changed to $\infty e^{i\alpha}$ with $|\alpha| \leq \pi - \delta$ show that the right-hand side is unaltered provided that $|\alpha + \mathrm{ph}\,z| \leq \pi/2 - \delta$ and $|\mathrm{ph}\,z| \leq \pi - \delta$.

7. If a_m is as in Exercise 1 and $\varphi(t)$ satisfied Eq.(3.5.2) with $|\alpha| \leq \pi - \delta$ prove that there is $m' < 1$ such that, for $|z| \geq m'$,

$$\int_0^{\infty e^{i\alpha}} e^{-zt} \varphi(t) dt = \sum_{m=0}^{\omega-1} \frac{a_m}{z^{m+1}} + (1 + \varkappa) \frac{a_\omega}{\omega! z^{\omega+1}} J(\omega, 1/z\omega)$$

subject to $|\alpha + \mathrm{ph}\,z| \leq \pi/2 - \delta$ and $|\mathrm{ph}\,z| \leq \pi - \delta$.

8. If a_m is as in Exercise 1 and $|f(te^{i\alpha})| < Ae^{\gamma t}$ for $t \geq 0$ with $|\alpha| \leq \pi - \delta$ prove that, when z is unlimited,

$$\int_0^{\infty e^{i\alpha}} e^{-zt} t^\mu f(t) = \sum_{m=0}^{n-1} \frac{(m+\mu)! a_m}{z^{m+\mu+1}} + \frac{(1 + \varkappa) a_n}{z^{n+\mu+1}} J(n + \mu, 1/z)$$

for $n \in \mathbf{N}$ provided that $|\alpha + \mathrm{ph}\,z| \leq \pi/2 - \delta$ and $|\mathrm{ph}\,z| \leq \pi - \delta$.

9. Let $|f(te^{i\alpha})| < Ae^{\gamma t}$ for $t \geq 0$ and $0 < \alpha < 2\pi$. Let $\mu > -1$ and $\nu \geq 1$. Show that, if ϵ is infinitesimal with $\mathrm{ph}\,\epsilon = \alpha$,

$$\int_0^{\infty} e^{-t^\nu} t^\mu f(\epsilon t) dt = \frac{1}{\nu} \sum_{m=0}^{n-1} \left(\frac{m+\mu+1}{\nu} - 1 \right)! a_m \epsilon^m + a_n \epsilon^n (1 + \varkappa) J(\nu, \mu + n, -\epsilon)$$

for $n \in \mathbf{N}$ where

$$J(\nu, \mu, \epsilon) = \int_0^{\infty} \frac{e^{-t^\nu} t^\mu}{1 + \epsilon t} dt.$$

Is there a generalisation to an integral of the type in Theorem 3.1.6?

10. If $|f(te^{i\alpha})| < At^p + B$ for $t \geq 0$, $0 < \alpha < 2\pi$ and ϵ is infinitesimal show that, for $n \in \mathbf{N}$,

$$\int_{-\infty+i\beta}^{\infty+i\beta} e^{-t^2/2} f(\epsilon e^t) dt = \sum_{m=0}^{n-1} (2\pi)^{\frac{1}{2}} a_m e^{m^2/2} \epsilon^m + a_n \epsilon^n (1+\varkappa) K(\beta, n, -\epsilon)$$

where $\mathrm{ph}\,\epsilon = \alpha - \beta$ and

$$K(\beta, \mu, \epsilon) = \int_{-\infty+i\beta}^{\infty+i\beta} \frac{e^{\mu t - t^2/2}}{1 + \epsilon e^t} dt.$$

Chapter 4
UNIFORM ASYMPTOTICS

4.1 A simple example

To introduce the topic of this chapter we consider

$$I(z) = \int_0^1 e^{-zt}dt = (1 - e^{-z})/z. \qquad (4.1.1)$$

As $|z| \to \infty$ with $|\text{ph } z| \le \pi/2 - \delta$

$$I(z) \sim 1/z \qquad (4.1.2)$$

but, as $\mathcal{R}(z) \to -\infty$,

$$I(z) \sim -e^{-z}/z. \qquad (4.1.3)$$

The two asymptotic forms are quite different and neither is valid when z tends to infinity along the imaginary axis. Here the full expression of Eq.(4.1.1) must be retained because both parts are bounded and of the same order of magnitude. Thus, the asymptotics of the function $I(z)$ depend crucially upon the value of ph z. As ph z increases through $\pi/2$, the approximation for $I(z)$ switches from Eq.(4.1.2) to Eq.(4.1.3) after passing through Eq.(4.1.1). This is an illustration of a *Stokes' phenomenon* since Stokes was the first to observe such behaviour. The imaginary axis, where the switch from one form to another takes place, is known as a *Stokes line*.

A formula like that of Eq.(4.1.1) which is valid on both sides of a Stokes line is said to be a *uniform* asymptotic representation, being uniformly valid as ph z varies. Usually, the transition across a Stokes line will be more complicated than in Eq.(4.1.1) but Eq.(4.1.1) serves as a reminder that behaviour in the neighbourhood of a Stokes line can be simple.

4.2 The function J

The function

$$J(\mu, \epsilon) = \int_0^\infty \frac{t^\mu e^{-t}}{1 + \epsilon t}dt$$

has occurred in several places in Chapter 3 and the approximation

$$J(\mu,\epsilon) = \mu!(1+\mu)/(1+\mu\epsilon) \qquad (4.2.1)$$

has been deployed when ϵ is not near the negative real axis. When ϵ crosses the negative real axis J changes discontinuously so that the negative real axis constitutes a branch line of J and its analytic continuation. The discontinuity is specified by

$$J(\mu,\epsilon) - J(\mu,\epsilon e^{-2\pi i}) = 2\pi i e^{\pi i\mu} e^{1/\epsilon}/\epsilon^{\mu+1}$$

or, more generally, by

$$J(\mu,\epsilon) - J(\mu,\epsilon e^{-2k\pi i}) = \frac{2\pi i}{\epsilon^{\mu+1}} e^{\mu k\pi i} \frac{\sin k\mu\pi}{\sin\mu\pi} e^{1/\epsilon} \qquad (4.2.2)$$

for integer k.

Sometimes it is more convenient to handle

$$J_0(\mu,z) = \int_0^\infty \frac{t^\mu e^{-t}}{t+z} dt = J(\mu,1/z)/z. \qquad (4.2.3)$$

From Eq.(4.2.2)

$$J_0(\mu,z) - J_0(\mu,ze^{-2k\pi i}) = -2\pi i e^{-\mu k\pi i} \frac{\sin k\mu\pi}{\sin\mu\pi} z^\mu e^z. \qquad (4.2.4)$$

There are no restrictions on Eq.(4.2.4) so long as $z=0$ is avoided.

These relations show that Eq.(4.2.1) is not an adequate description of J near the negative real axis. To cover this neighbourhood a formula is needed for J which is uniformly valid as ϵ varies near the negative real axis.

In fact, it is more convenient to discuss

$$J_1(\mu,z) = \int_0^\infty \frac{t^\mu e^{-t}}{t-z} dt = J_0(\mu,ze^{\pi i}) = -J(\mu,e^{-\pi i}/z)/z \qquad (4.2.5)$$

in which the uniform behaviour near the positive real axis is desired. If J_1 is determined for $|\mathrm{ph}\, z| < \pi$ then $J_0(\mu,z)$ is known on the sheet $0 < \mathrm{ph}\, z < 2\pi$ which includes the negative real axis.

For applications in asymptotics μ is generally large so the restriction $\mu > 0$ will be imposed.

Make the substitution

$$\mu \ln(t/\mu) - t + \mu = -\{r(t)\}^2/2\mu. \qquad (4.2.6)$$

To determine $r(t)$ the square root of the left-hand side has to be specified. The left-hand side increases from $-\infty$ to 0 as t goes from 0 to μ and then decreases to $-\infty$ as t increases to infinity. Therefore, take $r(t) < 0$ when $t < \mu$ and $r(t) > 0$ when $t > \mu$. With that convention the expansion of $r(t)$ around $t = \mu$ is

$$
\begin{aligned}
r(t) = \ & t - \mu - (t - \mu)^2/3\mu + 7(t - \mu)^3/36\mu^2 - 73(t - \mu)^4/540\mu^3 \\
& - 1331(t - \mu)^5/12960\mu^4 + O\{(t - \mu)^6\}.
\end{aligned} \tag{4.2.7}
$$

Then

$$
J_1(\mu, z) = e^\nu \int_{-\infty}^{\infty} \frac{e^{-r^2/2\mu}}{t - z} \frac{dt}{dr} dr
$$

where r has been written for $r(t)$ and

$$
\nu = \mu \ln \mu - \mu. \tag{4.2.8}
$$

The pole at $t = z$ can be accommodated by writing

$$
\frac{1}{t - z} \frac{dt}{dr} = \frac{B}{r - r(z)} + C + k(r) \tag{4.2.9}
$$

where $k(r)$ is a regular function of r such that $k(0) = 0$.

The determination of $r(z)$ from Eq.(4.2.6) is more elaborate than that of $r(t)$ because z is complex. We wish to arrange that $r(z) \to r(t)$ as $z \to t$ so that $r(z)$ is suitably continuous. The formula of Eq.(4.2.7) discloses that the imaginary part of $r(z)$ has the same sign as the imaginary part of z when z is near μ. If this is true for general z then the rule for fixing $r(z)$ is known—the sign of $\mathcal{I}\{r(z)\}$ has to be the same as the sign of $\mathcal{I}(z)$.

Suppose that ph $z = \theta$ and $r(z) = r_1 + ir_2$ where r_1 and r_2 are real. From Eq.(4.2.6)

$$
(r_1^2 - r_2^2)/2\mu = |z| \cos \theta - \mu - \mu \ln(|z|/\mu), \tag{4.2.10}
$$

$$
r_1 r_2 = \mu(|z| \sin \theta - \mu\theta). \tag{4.2.11}
$$

As $\theta \to 0$, $r_1 r_2 \to 0$ and $r_1^2 - r_2^2$ is positive so long as $|z| \neq \mu$. Therefore $r_2 \to 0$ i.e. r_2 is small when $|\theta|$ is small. Indeed, for small positive θ, $r_1 r_2$ has the same sign as $|z| - \mu$. To satisfy $r(z) \to r(t)$ as $z \to t$ we must have $r_1 > 0$ when $|z| > \mu$ and $r_1 < 0$ when $|z| < \mu$. Consequently, r_2 is positive for small positive θ. In other words, r_2 has the same sign as $\mathcal{I}(z)$ when z is just above the positive real axis. A repetition of the argument with θ negative reveals that this is still valid for z just below.

Now allow θ to grow to π; r_2 must remain positive unless it passes through a zero. Eq.(4.2.11) shows that this is impossible if $|z| < \mu$. When $|z| \geq \mu$, the right-hand side of Eq.(4.2.11) increases as θ increases until it reaches a maximum at $\theta = \theta_m$ where $\cos \theta_m = \mu/|z|$ and then decreases to a negative value at

$\theta = \pi$. Any zero occurs where $\theta > \theta_m$. But the right-hand side of Eq.(4.2.10) is a diminishing function of θ and is negative at $\theta = \theta_m$ since $|z| > \mu$. Hence r_2 cannot possibly vanish for $\theta > \theta_m$ and confirmation that r_2 does not change sign has been secured. A similar argument comes to the same conclusion when z is below the real axis.

Thus, the rule for choosing $r(z)$ is that its imaginary part should have the same sign as $\mathcal{I}(z)$.

The constant B in Eq.(4.2.9) must ensure that the singularity at $r = r(z)$ balances that at $t = z$. In terms of t

$$[r^2 - \{r(z)\}^2]/2\mu = \mu \ln(z/t) + t - z \approx (t - z)(1 - \mu/z)$$

so that $r - r(z) \approx \mu(t - z)(1 - \mu/z)/r$. Since $dt/dr = rt/\mu(t - \mu)$ we infer that $B = 1$. By putting $r = 0$ $(t = \mu)$ we see that

$$C = \frac{1}{r(z)} + \frac{1}{\mu - z}.$$

For $\mathcal{I}\{r(z)\} < 0$

$$
\begin{aligned}
\int_{-\infty}^{\infty} \frac{e^{-r^2/2\mu}}{r - r(z)} dr &= -i \int_{-\infty}^{\infty} e^{-r^2/2\mu} \int_0^{\infty} e^{iy\{r - r(z)\}} dy\, dr \\
&= -i(2\pi\mu)^{\frac{1}{2}} \int_0^{\infty} e^{-iyr(z) - \mu y^2/2} dy
\end{aligned}
$$

after interchanging the order of integration. The last integral can be expressed in terms of the complementary error function

$$\mathrm{erfc}(w) = \frac{2}{\pi^{\frac{1}{2}}} \int_w^{\infty} e^{-y^2} dy$$

with the result that

$$\int_{-\infty}^{\infty} \frac{e^{-r^2/2\mu}}{r - r(z)} dr = -\pi i e^{-\{r(z)\}^2/2\mu} \mathrm{erfc}\{ir(z)/(2\mu)^{\frac{1}{2}}\}.$$

For $\mathcal{I}\{r(z)\} > 0$ change the sign of i throughout and use the fact that

$$\mathrm{erfc}(-z) = 2 - \mathrm{erfc}(z).$$

The net effect is that

$$\int_{-\infty}^{\infty} \frac{e^{-r^2/2\mu}}{r - r(z)} dr = \pi i e^{-\{r(z)\}^2/2\mu} [2H\{\mathcal{I}(z)\} - \mathrm{erfc}\{ir(z)/(2\mu)^{\frac{1}{2}}\}] \qquad (4.2.12)$$

where $H(x)$ is the Heaviside step function introduced in Section 3.5.

It follows from Eqs.(4.2.6)–(4.2.9) and Eq.(4.2.12) that

$$J_1(\mu,z) = \pi i z^\mu e^{-z}[2H\{\mathcal{I}(z)\} - \text{erfc}\{ir(z)/(2\mu)^{\frac{1}{2}}\}]$$
$$+(2\pi\mu)^{\frac{1}{2}}e^\nu\left\{\frac{1}{r(z)} + \frac{1}{\mu-z}\right\} + e^\nu\int_{-\infty}^\infty k(r)e^{-r^2/2\mu}dr.$$

$$(4.2.13)$$

The final integral of Eq.(4.2.13) is expected to make a smaller contribution than the pole which has been accounted for already. An assessment of its influence can be arrived at by noting that $k(r)$ is regular and can be expanded via

$$k(r) = \sum_{p=1} A_p r^p.$$

Hence

$$\int_{-\infty}^\infty k(r)e^{-r^2/2\mu}dr = \sum_{p=1}(p-\tfrac{1}{2})!(2\mu)^{p+\frac{1}{2}}A_{2p}.$$

The coefficients A_p can be determined in a number of ways from Eq.(4.2.9). Derivatives with respect to r may be taken and then r made zero. The derivatives can be found analytically or by taking advantage of the symbolic facilities of MATHEMATICA or MAPLE. Another way of reaching the structure of the coefficients is to observe that there is no singularity in Eq.(4.2.9) as r and $r(z)$ vary. In particular, there is none when both r and $r(z)$ are zero. But the nth derivative of the term containing B will involve $\{1/r(z)\}^{n+1}$ when $r = 0$. A singularity would be created when $r(z) = 0$ unless it was cancelled by a corresponding singularity from the left-hand side. Hence the left-hand side must produce the singular part of the expansion of $\{1/r(z)\}^{n+1}$ in powers of $\mu - z$. Whichever method is adopted it is found that

$$A_2 = \left\{\frac{1}{r(z)}\right\}^3 + \frac{1}{(\mu-z)^3} - \frac{1}{\mu(\mu-z)^2} + \frac{1}{12\mu^2(\mu-z)}, \qquad (4.2.14)$$

$$A_4 = \left\{\frac{1}{r(z)}\right\}^5 + \frac{1}{(\mu-z)^5} - \frac{5}{3\mu(\mu-z)^4}$$
$$+\frac{25}{36\mu^2(\mu-z)^3} - \frac{1}{36\mu^3(\mu-z)^2} + \frac{1}{864\mu^4(\mu-z)}, \qquad (4.2.15)$$

$$A_6 = \left\{\frac{1}{r(z)}\right\}^7 + \frac{1}{(\mu-z)^7} - \frac{7}{3\mu(\mu-z)^6} + \frac{7}{4\mu^2(\mu-z)^5} - \frac{77}{180\mu^3(\mu-z)^4}$$
$$+\frac{49}{4320\mu^4(\mu-z)^3} - \frac{1}{4320\mu^5(\mu-z)^2} - \frac{139}{777600\mu^6(\mu-z)}. \qquad (4.2.16)$$

When z is near μ the forms of Eqs.(4.2.14)–(4.2.16) and the formula for C are not very suitable for computation because of the numerous cancellations which take place. Expansions for z near μ are

$$
\frac{1}{r(z)} + \frac{1}{\mu - z} \approx \frac{1}{3\mu} - \frac{z-\mu}{12\mu^2} + \frac{23}{540\mu^3}(z-\mu)^2 - \frac{353}{12960\mu^4}(z-\mu)^3
$$
$$
+ \frac{589}{30240\mu^5}(z-\mu)^4 - \frac{81083(z-\mu)^5}{5443200\mu^6} + \frac{7783(z-\mu)^6}{653184\mu^7},
$$

$$(4.2.17)$$

$$
A_2 \approx -\frac{1}{540\mu^3} - \frac{z-\mu}{288\mu^4} + \frac{23(z-\mu)^2}{6048\mu^5} - \frac{3733(z-\mu)^3}{1088640\mu^6}, \quad (4.2.18)
$$

$$
A_4 \approx -\frac{25}{18144\mu^5} + \frac{139(z-\mu)}{155520\mu^6} - \frac{259(z-\mu)^2}{466560\mu^7} + \frac{7717(z-\mu)^3}{22394880\mu^8},
$$

$$(4.2.19)$$

$$
A_6 \approx \frac{101}{2332800\mu^7} + \frac{571(z-\mu)}{37324800\mu^8} - \frac{2016373(z-\mu)^2}{5542732800\mu^9}
$$
$$
+ \frac{194036993(z-\mu)^3}{4655895552000\mu^{10}}.
$$

$$(4.2.20)$$

Notice that these formulae imply that A_{2p} is of the order of $1/\mu^{2p+1}$ whether $\mu - z$ is small or not. In so far as μ-dependence is concerned, the terms in the integral of $k(r)$ diminish like $1/\mu^{p+\frac{1}{2}}$ and so the series can be expected to furnish suitable asymptotic behaviour for large μ.

The final result is

$$
\begin{aligned}
J_1(\mu, z) =\ & \pi i z^\mu e^{-z} [2H\{\mathcal{I}(z)\} - \text{erfc}\{ir(z)/(2\mu)^{\frac{1}{2}}\}] \\
& + (2\pi\mu)^{\frac{1}{2}} e^\nu \left\{ \frac{1}{r(z)} + \frac{1}{\mu - z} \right\} \\
& + e^\nu \sum_{p=1} (p - \tfrac{1}{2})! (2\mu)^{p+\frac{1}{2}} A_{2p}.
\end{aligned}
$$

$$(4.2.21)$$

The purpose in deriving Eq.(4.2.21) was to provide an expression for J_1 which held uniformly for z near the positive real axis especially when μ is large. Actually, it is valid over a much wider region of complex z. Suppose that $\mathcal{I}(z) < 0$ and $\left| r(z)/(2\mu)^{\frac{1}{2}} \right| \gg 1$. Then the asymptotic expansion

$$
\text{erfc}\, w \sim \frac{e^{-w^2}}{\pi^{\frac{1}{2}} w} \left\{ 1 + \sum_{p=1}^\infty \frac{(p - \frac{1}{2})!(-)^p}{\pi^{\frac{1}{2}} w^{2p}} \right\} \qquad (|\text{ph}\, w| \le \pi/2 - \delta) \qquad (4.2.22)
$$

together with Eqs.(4.2.14)–(4.2.16) shows that all the terms involving $1/r(z)$ cancel and

$$
J_1(\mu, z) = \frac{(2\pi\mu)^{\frac{1}{2}} e^\nu}{\mu - z}(1 + \varkappa)
$$

for large μ. By means of Stirling's formula

$$J_1(\mu, z) = \mu!(1 + \varkappa)/(\mu - z)$$

which is the same as would be deduced from Eq.(4.2.1) and Eq.(4.2.5) when z is not close to the positive real axis. The same conclusion can be drawn when $\mathcal{I}(z) > 0$ by using $\mathrm{erfc}(-w) = 2 - \mathrm{erfc}(w)$.

Pertinent to the meaning of z^μ in Eq.(4.2.21) is that it has been assumed that $|\mathrm{ph}\, z| \leq \pi$ in the derivation. At first sight this makes Eq.(4.2.21) discontinuous as z approaches the negative real axis from above and below whereas $J_1(\mu, z)$ is obviously continuous there. However, the cancellation which has been described just above ensures that there is, in fact, no discontinuity. For ph z outside the range $(-\pi, \pi)$ it is best to employ Eq.(4.2.2) or Eq.(4.2.4) to get ph z back into the range before taking advantage of Eq.(4.2.21).

The more general integral

$$\int_0^\infty f(t) \frac{t^\mu e^{-t}}{t - z} dt$$

can be tackled in a similar fashion. Instead of Eq.(4.2.9)

$$\frac{f(t)}{t - z} \frac{dt}{dr} = \frac{f(z)}{r - r(z)} + \frac{f(z)}{r(z)} + \frac{f(\mu)}{\mu - z} + \sum_{p=1} B_p r^p$$

holds. Formulae for the coefficients of even index are

$$B_2 = f(z)A_2 + h_2(z) + \frac{h_1(z)}{\mu} + \frac{h_0(z)}{12\mu^2}, \tag{4.2.23}$$

$$B_4 = f(z)A_4 + h_4(z) + \frac{5}{3\mu}h_3(z) + \frac{25}{36\mu^2}h_2(z) + \frac{h_1(z)}{36\mu^3} + \frac{h_0(z)}{864\mu^4}, \tag{4.2.24}$$

$$B_6 = f(z)A_6 + h_6(z) + \frac{7}{3\mu}h_5(z) + \frac{7}{4\mu^2}h_4(z) + \frac{77h_3(z)}{180\mu^3}$$
$$+ \frac{49h_2(z)}{4320\mu^4} + \frac{h_1(z)}{4320\mu^5} - \frac{139h_0(z)}{777600\mu^6} \tag{4.2.25}$$

where A_2, A_4, A_6 are given by Eqs.(4.2.14)–(4.2.16) and

$$h_n(z) = \frac{1}{(z - \mu)^{n+1}} \left\{ f(z) - \sum_{p=0}^n \frac{(z - \mu)^p}{p!} f^{(p)}(\mu) \right\}. \tag{4.2.26}$$

An alternative means of specifying h_n is

$$f(\mu) = f(z) - (z - \mu)h_0(z), \quad f^{(n)}(\mu) = n!\{h_{n-1}(z) - (z - \mu)h_n(z)\}.$$

For z near μ

$$B_2 \approx -\frac{f(\mu)}{540\mu^3} + \frac{f'(\mu)}{12\mu^2} + \frac{f''(\mu)}{2\mu} + \frac{f'''(\mu)}{6}$$
$$+ \left\{ -\frac{f(\mu)}{288\mu^4} - \frac{f'(\mu)}{540\mu^3} + \frac{f''(\mu)}{24\mu^2} + \frac{f'''(\mu)}{6\mu} + \frac{f^{(4)}(\mu)}{24} \right\} (z-\mu),$$

$$(4.2.27)$$

$$B_4 \approx -\frac{25f(\mu)}{18144\mu^5} + \frac{f'(\mu)}{864\mu^4} + \frac{f''(\mu)}{72\mu^3} + \frac{25f'''(\mu)}{216\mu^2} + \frac{5f^{(4)}(\mu)}{72\mu}$$
$$+ \frac{f^{(5)}(\mu)}{120} + \{973f(\mu) - 1500\mu f'(\mu) + 630\mu^2 f''(\mu)$$
$$+ 5040\mu^3 f'''(\mu) + 31500\mu^4 f^{(4)}(\mu) + 15120\mu^5 f^{(5)}(\mu)$$
$$+ 1512\mu^6 f^{(6)}(\mu)\} \frac{z-\mu}{1088640\mu^6},$$

$$(4.2.28)$$

$$B_6 \approx \{707f(\mu) - 2919\mu f'(\mu) + 1890\mu^2 f''(\mu) + 30870\mu^3 f'''(\mu)$$
$$+ 291060\mu^4 f^{(4)}(\mu) + 238140\mu^5 f^{(5)}(\mu) + 52920\mu^6 f^{(6)}(\mu)$$
$$+ 3240\mu^7 f^{(7)}(\mu)\}/16329600\mu^7 + \{3997f(\mu) + 11312\mu f'(\mu)$$
$$- 23352\mu^2 f''(\mu) + 10080\mu^3 f'''(\mu) + 123480\mu^4 f^{(4)}(\mu)$$
$$+ 931392\mu^5 f^{(5)}(\mu) + 635040\mu^6 f^{(6)}(\mu) + 120960\mu^7 f^{(7)}(\mu)$$
$$+ 6480\mu^8 f^{(8)}(\mu)\}(z-\mu)/261273600\mu^8,$$

$$(4.2.29)$$

$$\frac{f(z)}{r(z)} + \frac{f(\mu)}{\mu-z} \approx \frac{f(\mu)}{3\mu} + f'(\mu) + \left\{ -\frac{f(\mu)}{12\mu^2} + \frac{f'(\mu)}{3\mu} + \frac{f''(\mu)}{2} \right\}(z-\mu)$$
$$+ \left\{ \frac{23f(\mu)}{540\mu^3} - \frac{f'(\mu)}{12\mu^2} + \frac{f''(\mu)}{6\mu} + \frac{f'''(\mu)}{6} \right\}(z-\mu)^2. \quad (4.2.30)$$

The formula analogous to Eq.(4.2.21) is

$$\int_0^\infty f(t) \frac{t^\mu e^{-t}}{t-z} dt \sim \pi i f(z) z^\mu e^{-z} [2H\{\mathcal{I}(z)\} - \text{erfc}\{ir(z)/(2\mu)^{\frac{1}{2}}\}]$$
$$+ (2\pi\mu)^{\frac{1}{2}} e^\nu \left\{ \frac{f(z)}{r(z)} + \frac{f(\mu)}{\mu-z} \right\}$$
$$+ e^\nu \sum_{p=1} (p - \tfrac{1}{2})! (2\mu)^{p+\frac{1}{2}} B_{2p}. \quad (4.2.31)$$

Sometimes it is more convenient to replace the erfc by another representation, namely

$$\text{erfc}(iw) = \frac{2}{\pi^{\frac{1}{2}}} e^{w^2 + \pi i/4} F(w e^{\pi i/4})$$

where

$$F(w) = e^{iw^2} \int_w^\infty e^{-it^2} dt. \quad (4.2.32)$$

The function $F(w)$ has the property

$$F(-w) = \pi^{\frac{1}{2}} e^{iw^2 - \pi i/4} - F(w) \tag{4.2.33}$$

and, when $|w|$ is large with $\pi/2 > \text{ph}\, w > -\pi$,

$$F(w) = -\frac{i}{2w} + \frac{1}{4w^3} + O\left(\frac{1}{|w|^5}\right). \tag{4.2.34}$$

4.3 Pole near a saddle point

The asymptotic behaviour of an integral with a saddle point has been considered in Section 2.4. When the integrand contains a pole, which may be near the saddle point, further examination is necessary in order to acquire an answer which is uniformly valid. In Section 2.4 the saddle point lies on a complex contour and the essence of the method there is to convert the integral into one along the real axis. Here it will be assumed that any such transformation has been accomplished so that consideration can be limited to integration along the real axis. Of course, the contribution of any pole captured during the deformation must be included in the final result.

The integral to be considered is

$$I = \int_c^b \frac{f(t)}{t - p} e^{-xh(t)} dt$$

where c and b are real. It will be supposed that x is an unlimited positive number and that $h(t)$ has a saddle point at $t = a$ where a is not near c or b. Further, $h(t) - h(a)$ will be taken to be positive when $t \neq a$ and bounded away from zero at the endpoints c and b. The function $f(t)$ will be assumed to be regular and the pole $t = p$ does not lie in the interval (c, b) though it may be close to a. Our goal will be confined to finding out how the contribution of the saddle point is affected by the presence of the pole.

Make the change of variable

$$h(t) - h(a) = \tfrac{1}{2} A u^2$$

where $A = h''(a)$, is positive and not infinitesimal. Take $u > 0$ when $t > a$ and $u < 0$ when $t < a$ so that $u \approx t - a$ when t is near a. Then, as in Section 2.4, the dominant contribution to I is

$$I \sim e^{-xh(a)} \int_{-\infty}^{\infty} \frac{f(t) A u}{(t - p) h'(t)} \exp(-\tfrac{1}{2} A x u^2) du$$

where here $\sim$ signifies that smaller terms are being neglected. As in Section 4.2 put

$$\frac{f(t)Au}{(t-p)h'(t)} = \frac{B}{u-u_p} + C + k(u) \tag{4.3.1}$$

where

$$\tfrac{1}{2}Au_p^2 = h(p) - h(a)$$

on the understanding that $u_p \approx p - a$ when p is close to a. This makes $\mathcal{I}(u_p)$ have the same sign as $\mathcal{I}(p)$ when p is near a but, unlike Section 4.2, we cannot infer that this is true for other positions of p without more detailed information on $h(t)$.

From Eq.(4.3.1)

$$B = f(p), \; C = \frac{f(p)}{u_p} + \frac{f(a)}{a-p}.$$

The function $k(u)$ will be ignored from now on because the analysis of Section 4.2 indicates that it will be of lesser importance than the terms which are retained. Hence

$$\begin{aligned} I \; \sim \; & \pi i f(p)e^{-xh(p)}[2H\{\mathcal{I}(u_p)\} - \mathrm{erfc}\{iu_p(Ax/2)^{\frac{1}{2}}\}] \\ & + \left\{\frac{f(p)}{u_p} + \frac{f(a)}{a-p}\right\}\left(\frac{2\pi}{Ax}\right)^{\frac{1}{2}} e^{-xh(a)} \end{aligned} \tag{4.3.2}$$

or

$$\begin{aligned} I \; \sim \; & 2\pi i f(p)e^{-xh(p)}H\{\mathcal{I}(u_p)\} - 2\pi^{\frac{1}{2}}if(p)e^{\pi i/4 - xh(a)}F\{e^{\pi i/4}u_p(Ax/2)^{\frac{1}{2}}\} \\ & + \left\{\frac{f(p)}{u_p} + \frac{f(a)}{a-p}\right\}\left(\frac{2\pi}{Ax}\right)^{\frac{1}{2}} e^{-xh(a)} \end{aligned} \tag{4.3.3}$$

in the notation of Eq.(4.2.32).

When p is not near a, u_p is not small. Therefore, the largeness of Ax entails, via Eq.(4.2.34) and Eq.(4.2.33),

$$I \sim \frac{f(a)}{a-p}\left(\frac{2\pi}{Ax}\right)^{\frac{1}{2}} e^{-xh(a)}.$$

This is the same result as in Example 2.4.2 of Chapter 2 for the contribution of a saddle point when there is no pole nearby. Thus, as p moves about Eq.(4.3.2) and Eq.(4.3.3) provide a smooth transition from the behaviour of a saddle point alone to the combined effect of saddle point and pole i.e. it is uniformly valid.

4.4   Saddle point near an endpoint

The discussion of the saddle point in Section 2.4 was based on the saddle point being either at an endpoint of the range of integration or well away from the endpoint. It is time now to consider what happens when the saddle point is in the vicinity of an endpoint. A typical case is the integral

$$I = \int_0^\infty f(t)t^\mu e^{-x^2 h(t)} dt$$

where x is positive and unlimited.

It will be assumed that $\mu > -1$ and that f is differentiable as many times as desired. The function $h(t)$ is supposed to be defined for negative values of t as well as positive and to possess a single minimum at $t = a$ where $h''(a) > 0$ but is not infinitesimal. The value of a may be positive or negative to allow for the saddle point being inside or outside the interval of integration. In addition, the condition that $h(t) \to \infty$ as $t \to \infty$ will be imposed.

Make the transformation

$$h(t) - h(a) = \tfrac{1}{2}(u - b)^2$$

so that $u = b$ corresponds to the saddle point $t = a$. For $t > a$ choose $u > b$ while, for $t < a$, $u < b$. Then

$$u - b \approx \{h''(a)\}^{\frac{1}{2}}(t - a)$$

when t is near a. With regard to b it is selected to have the same sign as a and to make $u = 0$ correspond to $t = 0$. Hence

$$b = \pm[2\{h(0) - h(a)\}]^{\frac{1}{2}}$$

according as $a \gtrless 0$.

After the transformation

$$I e^{x^2 h(a)} = \int_0^\infty u^\mu g_0(u) \exp\{-\tfrac{1}{2}x^2(u - b)^2\} du$$

where

$$g_0(u) = f(t)\left(\frac{t}{u}\right)^\mu \frac{dt}{du}.$$

The target now is to expand $g_0(u)$ near the origin and express quantities in terms of

$$\int_0^\infty u^\mu \exp\{-\tfrac{1}{2}x^2(u - b)^2\} du = V_\mu(bx)/x^{\mu+1}$$

where

$$V_\mu(x) = \int_0^\infty v^\mu e^{-\frac{1}{2}(v - x)^2} dv.$$

The function V_μ can be rewritten as a parabolic cylinder function. There are two notations for the parabolic cylinder function $U(\nu, x)$ and $D_\nu(x)$ (Abramowitz & Stegun 1965) in terms of which

$$V_\mu(x) = \mu! e^{-x^2/4} D_{-\mu-1}(-x) = \mu! e^{-x^2/4} U(\mu + \tfrac{1}{2}, -x). \qquad (4.4.1)$$

To enable the integral to be expressed via V_μ put

$$g_0(u) = \alpha_0 + \beta_0(u - b) + u(u - b)G_0(u).$$

Then

$$
\begin{aligned}
\alpha_0 &= g_0(b) = f(a)\left(\frac{a}{b}\right)^\mu \Big/ \{h''(a)\}^{\frac{1}{2}}, \\
\beta_0 &= \{g_0(b) - g_0(0)\}/b = \left[g_0(b) - f(0)\left\{-\frac{b}{h'(0)}\right\}^{\mu+1}\right]\Big/ b.
\end{aligned}
$$

There is no difficulty over the interpretation of $g_0(0)$ because b always has the opposite sign to $h'(0)$ on account of the conventions adopted. Now

$$I e^{x^2 h(a)} = \frac{\alpha_0}{x^{\mu+1}} V_\mu(bx) + \frac{\beta_0}{x^{\mu+2}} V'_\mu(bx) + \int_0^\infty u^{\mu+1}(u - b)G_0(u)e^{-\frac{1}{2}x^2(u-b)^2} du. \qquad (4.4.2)$$

Integration by parts yields

$$\int_0^\infty u^{\mu+1}(u - b)G_0(u)e^{-\frac{1}{2}x^2(u-b)^2} du = \frac{1}{x^2}\int_0^\infty u^\mu g_1(u)e^{-\frac{1}{2}x^2(u-b)^2} du$$

with

$$g_1(u) = (\mu + 1)G_0(u) + uG'_0(u).$$

Treat g_1 in the same way as g_0 i.e. write

$$g_1(u) = \alpha_1 + \beta_1(u - b) + u(u - b)G_1(u).$$

Clearly, the effect is to add α_1/x^2 to α_0 and β_1/x^2 to β_0. The procedure can be repeated and generates an asymptotic series in powers of $1/x^2$ multiplying $V_\mu(bx)$ and a similar one as a factor of $V'_\mu(bx)$.

Evidently, the dominant part of the asymptotic expansion is given by the first two terms of Eq.(4.4.2) i.e.

$$I \sim \left\{\frac{\alpha_0}{x^{\mu+1}} V_\mu(bx) + \frac{\beta_0}{x^{\mu+2}} V'_\mu(bx)\right\} e^{-x^2 h(a)} \qquad (4.4.3)$$

to a first approximation. This approximation varies continuously as the saddle point moves through the endpoint and its performance should be checked against earlier results for the saddle point in coincidence with the endpoint and well away from it.

When the saddle point coincides with the endpoint $a = 0$ and $b = 0$. Either directly from the integral for V_μ or from Eq.(4.4.1) and the known values for parabolic cylinder functions (Abramowitz & Stegun 1965)

$$V_\mu(0) = (\tfrac{1}{2}\mu - \tfrac{1}{2})! 2^{\frac{1}{2}\mu - \frac{1}{2}}, \; V_\mu'(0) = (\tfrac{1}{2}\mu)!(-2^{\frac{1}{2}\mu}) \tag{4.4.4}$$

on deploying $\mu! = (\tfrac{1}{2}\mu)!(\tfrac{1}{2}\mu - \tfrac{1}{2})! 2^\mu / \pi^{\frac{1}{2}}$. Hence

$$\begin{aligned} I \; &\sim \; (\tfrac{1}{2}\mu - \tfrac{1}{2})! 2^{\frac{1}{2}\mu - \frac{1}{2}} \alpha_0 e^{-x^2 h(0)} / x^{\mu+1} \\ &\sim \; (\tfrac{1}{2}\mu - \tfrac{1}{2})! \frac{f(0)}{2} \left\{ \frac{2}{h''(0)x^2} \right\}^{\frac{1}{2}(\mu+1)} e^{-x^2 h(0)} \end{aligned}$$

since b/a tends to $\{h''(0)\}^{\frac{1}{2}}$ in the limit. Agreement with Eq.(2.3.1) of Chapter 2 is confirmed on making the appropriate changes of parameters there.

For the saddle point away from the endpoint the separation will be regarded as sufficient for $|bx| \gg 1$. Asymptotic formulae for V_μ can be introduced then. They are

$$V_\mu(x) \sim (2\pi)^{\frac{1}{2}} x^\mu, \; V_\mu'(x) \sim (2\pi)^{\frac{1}{2}} \mu x^{\mu-1} \tag{4.4.5}$$

as $x \to \infty$, and

$$V_\mu(x) \sim \mu! e^{-\frac{1}{2}x^2} (-x)^{-\mu-1}, \; V_\mu'(x) \sim \mu! e^{-\frac{1}{2}x^2} (-x)^{-\mu} \tag{4.4.6}$$

as $x \to -\infty$. When $a > 0$, $b > 0$ and Eq.(4.4.5) is relevant; thus

$$I \sim \frac{\alpha_0}{x}(2\pi)^{\frac{1}{2}} b^\mu e^{-x^2 h(a)} \sim \left\{ \frac{2\pi}{h''(a)} \right\}^{\frac{1}{2}} f(a) \frac{a^\mu}{x} e^{-x^2 h(a)}$$

which is consistent with the result of Section 2.4 for the contribution of an interior saddle point (cf. Eq.(2.4.5) and Example 2.4.2 there).

When the saddle point is outside the interval of integration both a and b are negative; then Eq.(4.4.6) is pertinent and

$$\begin{aligned} I \; &\sim \; \frac{\mu!(\alpha_0 - b\beta_0)}{x^{2\mu+2}(-b)^{\mu+1}} e^{-x^2 h(0)} \\ &\sim \; \mu! f(0) \left\{ \frac{1}{h'(0)x^2} \right\}^{\mu+1} e^{-x^2 h(0)} \end{aligned}$$

which is in harmony with Eq.(2.3.1) of Chapter 2 when there is no saddle point in the range of integration.

Consequently, as the saddle point moves from outside to inside the interval of integration, Eq.(4.4.3) provides a continuous transition between values that have been secured before. Once again, a uniformly valid result has been achieved.

4.5   Coalescing saddle points

As already remarked the contribution of an isolated saddle point has been dealt with in Section 2.4. The presence of other saddle points which may not be far away produces a more complicated situation. A typical example is furnished by

$$\int f(w)e^{-xh(w,\sigma)}dw$$

in which x is unlimited positive and h depends upon the parameter σ. As σ varies a saddle point moves about and, on departing from one position to another, its influence on the integral may wax or wane. The possibility that it may pass through another saddle point must be taken into account. Suppose, for instance, that

$$h(w,\sigma) = \tfrac{1}{3}w^3 - \sigma^2 w. \qquad (4.5.1)$$

If $\sigma \neq 0$ there are two saddle points $w = \pm\sigma$ and they are of the second order. If $|\sigma|$ is large enough the contribution of each can be assessed separately by means of Section 2.4. However, as $|\sigma|$ reduces, they get closer and closer together until, when $\sigma = 0$, they coalesce into a single saddle point at $w = 0$ which is of the third order. Since second and third order saddle points generate quite different asymptotic contributions the need for a suitable transitional formula is evident.

   More generally, let $h(w,\sigma)$ possess two saddle points of the second order which coalesce into a single saddle point of the third order when $\sigma = 0$. To be specific let the solutions of $\partial h(w,\sigma)/\partial w = 0$ be $w_1(\sigma)$ and $w_2(\sigma)$ where $w_1(\sigma) = w_2(\sigma)$ when $\sigma = 0$ but otherwise w_1 and w_2 differ. The form of Eq.(4.5.1) models this kind of conduct which suggests the introduction of a new variable u via

$$h(w,\sigma) = \tfrac{1}{3}u^3 - \zeta(\sigma)u + g(\sigma) \qquad (4.5.2)$$

where

$$
\begin{aligned}
g(\sigma) &= \tfrac{1}{2}\{h(w_1,\sigma) + h(w_2,\sigma)\}, \\
\zeta^{3/2}(\sigma) &= \tfrac{3}{4}\{h(w_2,\sigma) - h(w_1,\sigma)\}.
\end{aligned}
$$

The insertion of $u = \zeta^{\frac{1}{2}}(\sigma)$ into Eq.(4.5.2) discloses that the corresponding value of w is $w_1(\sigma)$. Likewise, $w = w_2(\sigma)$ corresponds to $u = -\zeta^{\frac{1}{2}}(\sigma)$. It was shown by Chester, Friedman & Ursell (1957) that the transformation of Eq.(4.5.2) makes u uniformly regular for small w and σ. The relation between u and w can be regarded as one-to-one for σ in the circle $|\sigma| \leq \Sigma$ and u in the circle $|u| \leq U$ which contains the image of the circle $|w| \leq W$. Remark that these properties are totally independent of x; they are attributes of the mapping of Eq.(4.5.2) alone.

On account of these properties, $f(w)dw/du$ can be expanded in a power series of u near $u = 0$. Rather than deploying a direct expansion we shall follow the procedure of Section 4.4 and write

$$f(w)\frac{dw}{du} = g_0(u,\sigma) = \alpha_0(\sigma) + u\beta_0(\sigma) + \{u^2 - \zeta(\sigma)\}G_0(u,\sigma). \qquad (4.5.3)$$

The last term in Eq.(4.5.3) vanishes at the saddle points and so is expected to offer a less significant contribution than the other two terms. The quantities $\alpha_0(\sigma)$ and $\beta_0(\sigma)$ are fixed by choosing $u = \pm\zeta^{\frac{1}{2}}$ with the result

$$\alpha_0(\sigma) = \tfrac{1}{2}\{g_0(\zeta^{\frac{1}{2}},\sigma) + g_0(-\zeta^{\frac{1}{2}},\sigma)\},$$
$$\beta_0(\sigma) = \tfrac{1}{2}\{g_0(\zeta^{\frac{1}{2}},\sigma) - g_0(-\zeta^{\frac{1}{2}},\sigma)\}/\zeta^{\frac{1}{2}}.$$

Then G_0 is known and will have an expansion in powers of u near $u = 0$. Note that in the calculation of g_0

$$\frac{dw}{du} = \left\{2\zeta^{\frac{1}{2}} \bigg/ \frac{\partial^2 h(w_1,\sigma)}{\partial w^2}\right\}^{\frac{1}{2}}$$

when $u = \zeta^{\frac{1}{2}}$ and $\sigma \neq 0$; for $u = -\zeta^{\frac{1}{2}}$ change the sign of $\zeta^{\frac{1}{2}}$ and replace w_1 by w_2. When $\sigma = 0$

$$\frac{dw}{du} = \left\{2 \bigg/ \frac{\partial^3 h(w_1,\sigma)}{\partial w^3}\right\}^{\frac{1}{3}}$$

when $u = 0$.

The mapping of Eq.(4.5.2) creates a new contour of integration. To fix ideas we will assume that it can be made the path of integration in Fig. 3.1.1. Then

$$\int f(w)e^{-xh(w,\sigma)}dw = 2\pi i e^{-xg(\sigma)}\left[\frac{\alpha_0(\sigma)}{x^{\frac{1}{3}}}\,\mathrm{Ai}(x^{\frac{2}{3}}\zeta) + \frac{\beta_0(\sigma)}{x^{\frac{2}{3}}}\,\mathrm{Ai}\,'(x^{\frac{2}{3}}\zeta)\right]$$
$$+e^{-xg(\sigma)}\int (u^2 - \zeta)G_0(u,\zeta)e^{-x(\frac{1}{3}u^3 - \zeta u)}du. \qquad (4.5.4)$$

The integral can be converted by an integration by parts into one of the type already considered and the process can be repeated. Since the integration by parts divides by x it is transparent that the major contribution to the asymptotic development comes from the first two terms of Eq.(4.5.4).

The definition of $\zeta^{3/2}(\sigma)$ does not specify $\zeta^{\frac{1}{2}}(\sigma)$ uniquely; three possibilities are open to it. Each of the choices can be expected to lead to a different contour of integration in the u-plane. Only one of these should be the contour selected in the preceding paragraph. To identify the corresponding $\zeta^{\frac{1}{2}}(\sigma)$ the argument runs as follows. Let $u = 0$ entail $w = w_0$ in Eq.(4.5.2). Let the point w near w_0 map into u neighbouring the origin. Then, if the line segment from w_0 to w

points at the original path of integration we want the segment from the origin to u to point at the selected contour i.e.

$$2\pi/3 < 2n\pi + \mathrm{ph}\, u < 4\pi/3$$

for some integer n. But, in a conformal mapping, a line segment is rotated positively through an angle $\mathrm{ph}\, du/dw$. Hence, from Eq.(4.5.2)

$$-\frac{\pi}{3} > \mathrm{ph}\, \zeta^{\frac{1}{2}}(\sigma) - \tfrac{1}{2}\,\mathrm{ph}(w - w_0) - \tfrac{1}{2}\,\mathrm{ph}\,\frac{\partial h(w_0, \sigma)}{\partial w} + n\pi > -\frac{2\pi}{3} \qquad (4.5.5)$$

for some integer n. These inequalities are sufficient to fix the phase of $\zeta^{\frac{1}{2}}(\sigma)$.

An interpretation of Eq.(4.5.5) is that, if the original path were deformed to pass through $w = w_0$, the contour in the u-plane would go through $u = 0$ and a tangent there would lie in the sectors $(2\pi/3, 4\pi/3)$ and $(-\pi/3, \pi/3)$.

Example 4.5.1

Consider

$$\int_0^\pi e^{-xh(w,\sigma)} dw$$

where

$$h(w, \sigma) = -i\{\cos w + (w - \tfrac{1}{2}\pi)\sin \sigma\}$$

with $0 < \delta \le \sigma \le \pi/2$. The equation for the saddle points is $\sin w = \sin \sigma$ with solutions $w = \sigma$ or $\pi - \sigma$. Then

$$\tfrac{2}{3}\zeta^{3/2}(\sigma) = i\left\{\cos \sigma + (\sigma - \tfrac{1}{2}\pi)\sin \sigma\right\}.$$

Since the quantity in $\{\ \}$ is positive the choices for $\mathrm{ph}\, \zeta^{\frac{1}{2}}$ are $\pi/6$, $5\pi/6$ and $3\pi/2$. Now $w_0 = \pi/2$ and

$$\frac{\partial h(w, \sigma)}{\partial w} = i(1 - \sin \sigma)$$

when $w = w_0$. In implementing Eq.(4.5.5) w can be taken real and greater than $\pi/2$ so that

$$-\frac{\pi}{12} > \mathrm{ph}\, \zeta^{\frac{1}{2}}(\sigma) + n\pi > -\frac{5\pi}{12}$$

which can be satisfied only by $\mathrm{ph}\, \zeta^{\frac{1}{2}}(\sigma) = 5\pi/6$ with $n = -1$. Thus

$$\zeta^{\frac{1}{2}}(\sigma) = e^{5\pi i/6}[3\{\cos \sigma + (\sigma - \tfrac{1}{2}\pi)\sin \sigma\}/2]^{1/3}.$$

For another disposition of the saddle points take

$$h(w, \sigma) = -i\{\cos w + (w - \tfrac{1}{2}\pi)\cosh \sigma\}$$

with $\sigma \geq 0$. In this case there are saddle points at $w = \frac{1}{2}\pi \pm i\sigma$ and

$$\tfrac{2}{3}\zeta^{3/2}(\sigma) = -\{\sigma \cosh \sigma - \sinh \sigma\}.$$

Again the quantity in $\{\ \}$ is positive so that $\mathrm{ph}\,\zeta^{\frac{1}{2}}(\sigma)$ can be $\pi/3$, π or $5\pi/3$. Here $\partial h/\partial w = -i(\cosh \sigma - 1)$ when $w = \pi/2$. Consequently Eq.(4.5.5) becomes

$$-\frac{7\pi}{12} > \mathrm{ph}\,\zeta^{\frac{1}{2}}(\sigma) + n\pi > -\frac{11\pi}{12}$$

which is satisfied by $\mathrm{ph}\,\zeta^{\frac{1}{2}}(\sigma) = \pi/3$ with $n = -1$. Accordingly

$$\zeta^{\frac{1}{2}}(\sigma) = e^{\pi i/3}[3(\sigma \cosh \sigma - \sinh \sigma)/2]^{1/3}.$$

When there are more than two saddle points σ should be replaced by the vector $\boldsymbol{\sigma}$ and m saddle points are permitted. The obvious analogue of Eq.(4.5.2) is the mapping

$$h(w, \boldsymbol{\sigma}) = \frac{1}{m+1}u^{m+1} + x_1 u + x_2 u^2 + \cdots + x_{m-1}u^{m-1} + g(\boldsymbol{\sigma}). \qquad (4.5.6)$$

Clearly, the analysis will end up with integrals in which the exponent consists of the right-hand side of Eq.(4.5.6) with $g(\boldsymbol{\sigma})$ absent. Properties of such integrals are hard to come by except when $m = 3$ (when they are called Pearcey integrals) and when $m = 4$ (swallowtail integrals). Therefore the matter of multiple coalescence of saddle points will not be pursued further.

Exercises on Chapter 4

1. Since

$$\frac{\partial}{\partial z} J_1(\mu, z) = \int_0^\infty \frac{t^\mu e^{-t}}{(t-z)^2}dt$$

when z is not positive real, a derivative of Eq.(4.2.18) could supply an asymptotic formula for the integral on the right. Can the formula be justified?

2. Estimate the asymptotic behaviour of

$$\int_0^\infty \frac{t^\mu \ln t}{t - z}e^{-t}dt.$$

3. Show that, to a first approximation,

$$
\begin{aligned}
e^z \int_0^1 t^\lambda e^{-zt}dt \ \sim\ & \frac{\lambda! e^z}{z^{\lambda+1}} - \sum_{m=0}^{M-1}\frac{\lambda!}{(\lambda - m)!\,z^{m+1}} \\
& + \lambda!(1 - e^{2\pi i\lambda})e^z\,\mathrm{erfc}\{ir(ze^{-\pi i})/(2\mu)^{\frac{1}{2}}\} \\
& + \frac{(-)^M e^\lambda (2\pi\mu)^{\frac{1}{2}}}{(-1-\lambda)!\,z^M}\left\{\frac{1}{\mu + z} + \frac{1}{r(ze^{-\pi i})}\right\}
\end{aligned}
$$

where $\mu = M - 1 - \lambda$.

   4. Show that, if $0 \le a \le \pi/2 - \delta$ with $\delta > 0$,

$$\int_0^{\pi/2} t^\mu e^{x^2(\cos t + t \sin a)} dt$$

$$\sim \left\{ \frac{\alpha_0}{x^{\mu+1}} V_\mu(bx) + \frac{\beta_0}{x^{\mu+2}} V_\mu'(bx) \right\} \exp\{x^2(\cos a + a \sin a)\}$$

where $b = \{2(\cos a + a \sin a - 1)\}^{\frac{1}{2}}$, $\alpha_0 = (a/b)^\mu (\cos a)^{-\frac{1}{2}}$ and

$$\beta_0 = \left\{ \left(\frac{a}{b}\right)^\mu \frac{1}{(\cos a)^{\frac{1}{2}}} - \left(\frac{b}{\sin a}\right)^{\mu+1} \right\} \bigg/ b.$$

   5. If $0 < \delta \le c \le 1$ show that

$$\int_0^1 \exp[x^2\{ct + \ln(1-t)\}]dt \sim \frac{1}{cx} \left(\frac{\pi}{2}\right)^{\frac{1}{2}} \mathrm{erfc}(bx/2^{\frac{1}{2}})$$

where $b = \{2(c-1-\ln c)\}^{\frac{1}{2}}$. What would be the next term in the approximation?

   6. If $\sigma \ge 0$ show that, to a first approximation,

$$\int_0^\infty \exp\{-x(\mathrm{sech}\,\sigma \sinh w - w)\}dw$$

$$\sim \frac{1}{x^{1/3}} \left(\frac{2\zeta^{\frac{1}{2}}}{\tanh \sigma}\right)^{\frac{1}{2}} \int_0^\infty \exp\left(-\frac{1}{3}u^3 + x^{2/3}\zeta u\right) du$$

where $\zeta^{\frac{1}{2}} = +\{3(\sigma - \tanh \sigma)/2\}^{1/3}$. What would the next stage of the approximation provide?

   7. The contour C is displayed in Fig. 4.5.1. If $0 < \sigma \le \pi/2$ show that

$$\frac{1}{\pi} \int_C \exp[ix\{\cos w + (w - \tfrac{1}{2}\pi)\sin \sigma\}]dw$$

$$\sim -\frac{2^{3/2} e^{-\pi i/3}}{x^{1/3}} \left(\frac{|\zeta^{\frac{1}{2}}|}{\cos \sigma}\right)^{\frac{1}{2}} \mathrm{Ai}(x^{2/3}\zeta)$$

where

$$\zeta^{\frac{1}{2}} = e^{5\pi i/6}[3\{\cos \sigma + (\sigma - \tfrac{1}{2}\pi)\sin \sigma\}/2]^{1/3}.$$

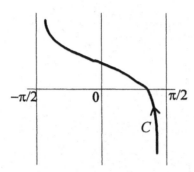

Figure 4.5.1 The contour for Exercises 7 and 8

8. With C the contour of Fig. 4.5.1 show that if $\sigma \geq 0$

$$\frac{1}{\pi} \int_C \exp[ix\{\cos w + (w - \tfrac{1}{2}\pi)\cosh\sigma\}]dw$$

$$\sim \quad -\frac{2^{3/2}e^{-\pi i/3}}{x^{1/3}} \left(\frac{|\zeta^{\frac{1}{2}}|}{\sinh\sigma} \right)^{\frac{1}{2}} \mathrm{Ai}(x^{2/3}\zeta)$$

where

$$\zeta^{\frac{1}{2}} = e^{\pi i/3}[3(\sigma\cosh\sigma - \sinh\sigma)/2]^{1/3}.$$

Chapter 5
HYPERASYMPTOTICS

5.1 Introduction

The general definition of an asymptotic expansion of a function $f(z)$ in terms of an asymptotic sequence $\{\varphi_n(z)\}$ was given in Chapter 1. It expresses $f(z)$ in the form

$$f(z) = \sum_{m=1}^{n} a_m \varphi_m(z) + R_n(z).$$

Generally, the series is divergent as $n \to \infty$ so that, to be of use, it has to be truncated at some value of n, say n_0, where the remainder $R_{n_0}(z)$ can be estimated or shown to be negligibly small. Usually an optimal n_0 is derived by minimising $R_n(z)$ for fixed z with the consequence that n_0 is a function of z generally. Chapter 3 had examples where the optimal remainder turned out to be exponentially small.

It was suggested by Stieltjes (1886) that the accuracy of the estimate for $f(z)$ could be improved further by expanding $R_{n_0}(z)$ itself in an asymptotic series. Then

$$f(z) = \sum_{m=1}^{n_0} a_m \varphi_m(z) + \sum_{m=1}^{n} a_{1m} \varphi_{1m}(z) + R_{1n}(z)$$

where the asymptotic sequence $\{\varphi_{1m}(z)\}$ need not be the same as $\{\varphi_m(z)\}$. The additional series is truncated at n_1 where $R_{1n_1}(z)$ is optimal and expected to be exponentially small compared with $R_{n_0}(z)$ as $|z| \to \infty$. Such a possibility was considered formally by Dingle (1973).

Of course, having carried out the process once we can repeat it i.e. expand $R_{1n_1}(z)$ in an asymptotic sequence up to its optimal remainder $R_{2n_2}(z)$ and so on. The continual repetition of this procedure was dubbed *hyperasymptotics* by Berry & Howls (1990, 1991) and is intended to supply steadily more accurate estimates of $f(z)$. The subject has been developed by several authors; see, for example, Berry (1991), Boyd (1990, 1993, 1994), Howls (1992), McLeod (1992), Olde Daalhuis (1992, 1993), Olde Daalhuis & Olver (1994), Olver (1991a, 1991b, 1994), Paris (1992a, 1992b), Paris & Wood (1992, 1995).

The question of determining the optimal $R_{n_0}(z)$ has been discussed in Chapter 3. Finding the optimal remainder of a remainder is a new topic and generally involves quite complicated manipulation. The following section illustrates this point.

5.2   A Laplace integral

The integral to be considered in this section is

$$I(x) = \int_0^\infty e^{-xt} t^\mu f(t) dt$$

where x is an unlimited positive number and the limited $\mu > -1$. It will be assumed that f possesses as many derivatives as are desired and that, for unlimited n,

$$\left| f^{(n)}(t) \right| \le \frac{(b+n)!K}{(p+t)^{n+1}} \quad (t \ge 0) \tag{5.2.1}$$

where K, b and p are standard constants independent of n with $p > 0$. Subsequently, K will be used as a generic standard constant, not necessarily the same in all places.

According to the theory already described for the asymptotics of a Laplace integral the dominant contribution is dictated by the behaviour of $f(t)$ near the origin. Therefore, make the expansion

$$f(t) = \sum_{m=0}^{n-1} \frac{t^m}{m!} f^{(m)}(0) + t^n f_1(t).$$

Then

$$I(x) = \sum_{m=0}^{n-1} \frac{(\mu+m)!}{m! x^{\mu+m+1}} f^{(m)}(0) + R_0(n,x) \tag{5.2.2}$$

where

$$R_0(n,x) = \int_0^\infty e^{-xt} t^{\mu+n} f_1(t) dt.$$

The first objective is to choose $n\,(= n_0)$ so that R_0 is exponentially small in x. Previous experience in Chapter 3 indicates that this will require n_0 to be of the order of x and unlimited. To fix n_0 necessitates some knowledge of the properties of $f_1(t)$. The usual formula for the remainder in a Maclaurin series is

$$f_1(t) = \frac{1}{(n-1)!} \int_0^1 (1-u)^{n-1} f^{(n)}(ut) dt.$$

Therefore

$$\left| f_1^{(r)}(t) \right| \le \frac{(b+n+r)!}{(n-1)!} K \int_0^1 \frac{(1-u)^{n-1} u^r}{(p+tu)^{n+r+1}} du$$

after calling on Eq.(5.2.1). The substitution $(p+q)u = (p+qu)v$ gives

$$\int_0^1 \frac{u^\mu(1-u)^{\nu-1}}{(p+qu)^{\mu+\nu+1}}du = \frac{\mu!(\nu-1)!}{(\mu+\nu)!p^\nu(p+q)^{\mu+1}} \tag{5.2.3}$$

for $q \geq 0$. Hence

$$\left|f_1^{(r)}(t)\right| \leq \frac{(b+n+r)!r!K}{(n+r)!p^n(p+t)^{r+1}}. \tag{5.2.4}$$

Accordingly

$$\begin{aligned}
|R_0(n,x)| &\leq \frac{(b+n)!K}{n!p^n}\int_0^\infty \frac{e^{-xt}t^{\mu+n}}{p+t}dt \\
&\leq \frac{(b+n)!K}{n!p^{n+1}}\int_0^\infty e^{-xt}t^{\mu+n}dt.
\end{aligned}$$

The last integral can be evaluated immediately as a factorial. However, as a guide to later analysis, it is better to estimate it asymptotically taking advantage of the unlimited parameters. In fact, it assists to replace the integer n by the parameter ν. Now

$$\int_0^\infty e^{-xt}t^{\mu+\nu}dt = \int_0^\infty \exp\{-xt + (\mu+\nu)\ln t\}dt.$$

The major contribution comes from a stationary point of the exponent; if the stationary point be denoted by t_1 it satisfies

$$-x + (\mu+\nu)/t_1 = 0. \tag{5.2.5}$$

Thus t_1 lies in the interval of integration and an expansion of the exponent around $t = t_1$ where the second derivative is $-(\mu+\nu)/t_1^2$ leads to

$$(2\pi)^{\frac{1}{2}}\exp[-xt_1 + (\mu+\nu)\ln t_1 + \tfrac{1}{2}\ln\{t_1^2/(\mu+\nu)\}]$$

as the main contribution to the integral. The implementation of Stirling's formula provides then

$$|R_0(\nu,x)| \leq K\exp\left\{b\ln\nu - (\nu+1)\ln p - xt_1 + (\mu+\nu)\ln t_1 + \tfrac{1}{2}\ln\frac{\mu+\nu}{x^2}\right\}$$

after using Eq.(5.2.5).

Allow ν to vary so as to achieve the smallest possible exponential; there will be a consequent variation of t_1 to meet Eq.(5.2.5). The changes are stationary when

$$b/\nu - \ln p + \ln t_1 + 1/2(\nu+\mu) = 0$$

by virtue of Eq.(5.2.5). Hence t_1 differs infinitesimally from p and it will be adequate to take

$$t_1 = p. \tag{5.2.6}$$

Then, from Eq.(5.2.5)
$$\nu = t_1 x - \mu = px - \mu.$$
However, in the end, we need n_0 to be an integer so choose
$$n_0 = [t_1 x - \mu] = [px - \mu] \tag{5.2.7}$$
where $[\nu]$ is the largest integer which does not exceed ν. Since p and μ are limited with p non-zero Eq.(5.2.7) does ensure that n_0 is unlimited; also
$$-x + (\mu + n_0)/t_1 \leq 0 \tag{5.2.8}$$
on account of Eq.(5.2.5).

With these choices of t_1 and n_0
$$|R_0(n_0, x)| \leq K(px)^{b-\frac{1}{2}} e^{-px}$$
showing that an exponentially small remainder has been attained.

To proceed to the next step we have to expand $R_0(n_0, x)$ until reaching an optimal remainder. Now it has been seen that the main contribution to R_0 comes from a neighbourhood of t_1. Therefore, we must expand $f_1(t)$ about this point i.e.
$$f_1(t) = \sum_{m=0}^{n-1} \frac{(t - t_1)^m}{m!} f_1^{(m)}(t_1) + (t - t_1)^n f_2(t).$$
Then
$$R_0(n_0, x) = \sum_{m=0}^{n-1} \frac{f_1^{(m)}(t_1)}{m!} \int_0^\infty e^{-xt} t^{\mu+n_0} (t - t_1)^m dt + R_1(n, x) \tag{5.2.9}$$
where
$$R_1(n, x) = \int_0^\infty e^{-xt} t^{\mu+n_0} (t - t_1)^n f_2(t) dt. \tag{5.2.10}$$
Here
$$f_2(t) = \frac{1}{(n-1)!} \int_0^1 (1 - u)^{n-1} f_1^{(n)}\{t_1 + u(t - t_1)\} du$$
from the usual Taylor expansion. Taking benefit from Eq.(5.2.4) and Eq.(5.2.3) we have
$$\left| f_2^{(r)}(t) \right| \leq \frac{(b + n_0 + n + r)!(n + r)!}{(n - 1)!(n_0 + n + r)!p^{n_0}} K \int_0^1 \frac{u^r (1 - u)^{n-1} du}{\{p + t_1 + u(t - t_1)\}^{n+r+1}}$$
$$\leq \frac{(b + n_0 + n + r)!r!K}{(n_0 + n + r)!p^{n_0}(p + t_1)^n (p + t)^{r+1}}. \tag{5.2.11}$$
Hence
$$|R_1(n, x)| \leq \frac{(b + n_0 + n)!K}{(n_0 + n)!p^{n_0+1}(p + t_1)^n} \int_0^\infty e^{-xt} t^{\mu+n_0} |t - t_1|^n \, dt.$$

Again, to minimise R_1 it is convenient to consider $R_1(\nu, x)$ with ν an unlimited parameter. The integrand has a stationary exponent at $t = t_2$ where

$$-x + (\mu + n_0)/t_2 + \nu/(t_2 - t_1) = 0 \qquad (5.2.12)$$

which will make ν positive by virtue of Eq.(5.2.8) so long as $t_2 > t_1$ as will be confirmed shortly. Then

$$
\begin{aligned}
|R_1(\nu, x)| \ \leq \ & K \exp\{b \ln(n_0 + \nu) - (n_0 + 1)\ln p - \nu \ln(p + t_1) - xt_2 \\
& + (\mu + n_0)\ln t_2 - \tfrac{1}{2}\ln s + \nu \ln(t_2 - t_1)\}
\end{aligned}
$$

where $s = (\mu + n_0)/t_2^2 + \nu/(t_2 - t_1)^2$. For variations in ν the exponent is stationary when

$$t_2 - t_1 = p + t_1$$

if infinitesimals are ignored. Therefore

$$t_2 = 3p = (2^2 - 1)p. \qquad (5.2.13)$$

With t_2 known, ν follows from Eq.(5.2.12) and a suitable integer n_1 for R_1 is $n_1 = [\nu]$. A consequence of Eq.(5.2.12) is that

$$-x + (\mu + n_0)/t_2 + n_1/(t_2 - t_1) \leq 0. \qquad (5.2.14)$$

The net result is that

$$|R_1(n_1, x)| \leq K(n_0 + n_1)^{b - \frac{1}{2}}\frac{t_2}{p} \exp\{-xt_2 + (\mu + n_0)\ln(t_2/t_1)\} \qquad (5.2.15)$$

since $s > (\mu + n_0 + n_1)/t_2^2$. The algebraic dependence on x is the same as in $R_0(n_0, x)$ since both n_0 and n_1 essentially contain x as a factor. As regards exponential behaviour that of Eq.(5.2.15) is smaller if

$$-x(t_2 - p) + (\mu + n_0)\ln(t_2/t_1) < 0.$$

Since

$$\ln(1 + y) < y - y^2/2(1 + y) \qquad (5.2.16)$$

for $y > 0$ the left-hand side is less than

$$
\begin{aligned}
& -x(t_2 - p) + (\mu + n_0)\left\{\frac{t_2 - t_1}{t_1} - \frac{(t_2 - t_1)^2}{2t_1 t_2}\right\} \\
< \ & -(\mu + n_0)(t_2 - t_1)^2/2t_1 t_2 \\
< \ & -2(\mu + n_0)/3
\end{aligned}
$$

on account of Eq.(5.2.6) and Eq.(5.2.8). Thus $R_1(n_1, x)$ does display a satisfactory smaller exponential behaviour than $R_0(n_0, x)$. In fact, if $\mu + n_0$ is replaced by px as a rough approximation in Eq.(5.2.15), the exponent is

$$px(-3 + \ln 3) = -1.90px$$

which is nearly double that in $R_0(n_0, x)$. Note also that, with the same approximation, n_1 is about $4(\mu + n_0)/3$ so that a third more terms are needed to reach R_1 from R_0 than are needed for R_0. Moreover, these terms involve the calculation of integrals which are no longer straightforward factorials. The price of increased accuracy is a substantial extra computational effort.

The next step is the expansion of $R_1(n_1, x)$ which entails expanding $f_2(t)$ about $t = t_2$. However, rather than do that, we will proceed to the general step under assumptions which are consistent with what has been found earlier. Then induction will validate the process.

Take

$$R_m(n, x) = \int_0^\infty e^{-xt} t^{\mu + n_0} (t - t_1)^{n_1} \ldots (t - t_{m-1})^{n_{m-1}} (t - t_m)^n f_{m+1}(t) dt \quad (5.2.17)$$

where

$$f_{m+1}(t) = \frac{1}{(n-1)!} \int_0^1 (1 - u)^{n-1} f_m^{(n)} \{t_m + u(t - t_m)\} du \qquad (5.2.18)$$

and

$$\left| f_m^{(r)}(t) \right| \le \frac{(b + n_0 + n_1 + \ldots + n_{m-1} + r)! r! K}{(n_0 + \ldots + n_{m-1} + r)! p^{n_0} (p + t_1)^{n_1} \ldots (p + t_{m-1})^{n_{m-1}} (p + t)^{r+1}}. \qquad (5.2.19)$$

Then, from Eq.(5.2.3)

$$\left| f_{m+1}^{(r)}(t) \right| \le \frac{(b + n_0 + n_1 + \ldots + n_{m-1} + n + r)! r! K}{(n_0 + \ldots + n_{m-1} + n + r)! p^{n_0} (p + t_1)^{n_1} \ldots (p + t_m)^n (p + t)^{r+1}}$$

which is the same as Eq.(5.2.19) with m altered to $m + 1$ when $n = n_m$. Hence

$$|R_m(\nu, x)|$$
$$\le \frac{(b + n_0 + n_1 + \ldots + n_{m-1} + \nu)! K}{(n_0 + \ldots + n_{m-1} + \nu)! p^{n_0 + 1} (p + t_1)^{n_1} \ldots (p + t_{m-1})^{n_{m-1}} (p + t_m)^\nu}$$
$$\times \int_0^\infty e^{-xt} t^{\mu + n_0} |t - t_1|^{n_1} \ldots |t - t_{m-1}|^{n_{m-1}} |t - t_m|^\nu dt.$$

The exponent of the integrand is stationary at $t = t_{m+1}$ where

$$-x + \frac{\mu + n_0}{t_{m+1}} + \frac{n_1}{t_{m+1} - t_1} + \cdots + \frac{n_{m-1}}{t_{m+1} - t_{m-1}} + \frac{\nu}{t_{m+1} - t_m} = 0 \qquad (5.2.20)$$

and then the exponent of R_m is stationary for

$$t_{m+1} - t_m = p + t_m \tag{5.2.21}$$

if infinitesimals are dropped. Eq.(5.2.21) is solved by

$$t_m = p(2^m - 1) \tag{5.2.22}$$

which matches Eq.(5.2.6) and Eq.(5.2.13). Then ν is known from Eq.(5.2.20) and we take $n_m = [\nu]$ so that

$$-x + \frac{\mu + n_0}{t_{m+1}} + \sum_{r=1}^{m} \frac{n_r}{t_{m+1} - t_r} \le 0. \tag{5.2.23}$$

On the other hand $n_m + 1 > [\nu]$ which implies

$$
\begin{aligned}
1 + n_m &> (t_{m+1} - t_m)\left\{ x - \frac{\mu + n_0}{t_{m+1}} - \sum_{r=1}^{m-1} \frac{n_r}{t_{m+1} - t_r} \right\} \\
&> (t_{m+1} - t_m)^2 \left\{ \frac{\mu + n_0}{t_m t_{m+1}} + \sum_{r=1}^{m-1} \frac{n_r}{(t_{m+1} - t_r)(t_m - t_r)} \right\}
\end{aligned} \tag{5.2.24}
$$

from Eq.(5.2.23) with $m - 1$ for m. It is evident that Eq.(5.2.24) ensures that

$$
\begin{aligned}
1 + n_m &> \frac{(t_{m+1} - t_m)^2 n_{m-1}}{(t_{m+1} - t_{m-1})(t_m - t_{m-1})} \\
&> 4n_{m-1}/3 \tag{5.2.25}
\end{aligned}
$$

by virtue of Eq.(5.2.22). Since n_m is unlimited one infers that essentially n_m grows at least as fast as a constant multiple of $(4/3)^m$ i.e. n_m is unbounded as $m \to \infty$.

Incorporating the above in R_m we have

$$
\begin{aligned}
|R_m(n_m, x)| &\le K \left(\sum_{r=0}^{m} n_r \right)^{b - \frac{1}{2}} \frac{t_{m+1}}{p} \exp\left\{ -x t_{m+1} + (\mu + n_0) \ln \frac{t_{m+1}}{t_1} \right. \\
&\left. + \sum_{r=1}^{m-1} n_r \ln \frac{t_{m+1} - t_r}{t_{r+1} - t_r} \right\}. \tag{5.2.26}
\end{aligned}
$$

It should be checked now that the exponent in Eq.(5.2.26) is less than that when m is replaced by $m - 1$. The difference between them is

$$
\begin{aligned}
&-x(t_{m+1} - t_m) + (\mu + n_0) \ln \frac{t_{m+1}}{t_m} + \sum_{r=1}^{m-1} n_r \ln \frac{t_{m+1} - t_r}{t_m - t_r} \\
&< -\frac{1}{2}(t_{m+1} - t_m)^2 \left\{ \frac{\mu + n_0}{t_m t_{m+1}} + \sum_{r=1}^{m-1} \frac{n_r}{(t_{m+1} - t_r)(t_m - t_r)} \right\}
\end{aligned}
$$

after employing Eq.(5.2.16) and Eq.(5.2.23). This is sufficient to show that further exponential improvement has been achieved. More can be said because Eq.(5.2.24) indicates that the difference is effectively less than $-\frac{1}{2}n_m$. Thus, the exponent in Eq.(5.2.26) is certainly less than $-\frac{1}{2}\sum_{r=1}^{m} n_r$. This quantity tends to $-\infty$ as $m \to \infty$ because n_m is unbounded. Consequently, the exponential can be made arbitrarily small by taking m large enough.

The numerical factor multiplying the exponential in Eq.(5.2.26) may grow or diminish as m increases depending on the sign of $b - \frac{1}{2}$. However, its effect is overwhelmed by the exponential decay. Accordingly, we conclude that *any desired level of accuracy can be attained by making m large enough*. Of course, the cost of obtaining this accuracy may be considerable in view of the large number of complicated terms which have to be computed.

It will be noticed that t_m, as given by Eq.(5.2.22), is unbounded as $m \to \infty$. The explanation lies in the nature of the successive expansions. In Eq.(5.2.2) the behaviour of the integrand in the neighbourhood of the origin has been taken care of effectively and R_0 has to account for the integrand elsewhere. The expansion of R_0 takes care of another neighbourhood leaving R_1 to look after the rest. The process continues and, since the upper limit of integration is infinite, neighbourhoods approaching infinity must occur eventually.

The preceding theory may be summarised briefly as:

expand successively each remainder about $t = t_m$ (given by Eq.(5.2.22)) to the number of terms specified by Eq.(5.2.20) until the remainder meets the relevant criterion of accuracy.

As an aid to the computation of the coefficients remark that

$$\int_0^\infty e^{-xt} t^{\mu+n_0}(t-t_1)^{n_1}\ldots(t-t_{m-1})^{n_{m-1}}(t-t_m)^r dt$$

$$= \sum_{s=0}^{r} \frac{r!(t_{m-1}-t_m)^{r-s}}{s!(r-s)!}$$

$$\times \int_0^\infty e^{-xt} t^{\mu+n_0}(t-t_1)^{n_1}\ldots(t-t_{m-2})^{n_{m-2}}(t-t_{m-1})^{n_{m-1}+s}dt$$

$$(5.2.27)$$

which expresses the integral on the left via an integral with one less factor in the integrand and thereby provides a recursion formula. The other quantity required $f_m^{(r)}(t_m)$ can be calculated recursively in two ways. Either the integral formula of Eq.(5.2.18) or the equivalent expansion

$$f_{m+1}(t) = \left\{ f_m(t) - \sum_{r=0}^{n_m-1} \frac{(t-t_m)^r}{r!} f_m^{(r)}(t_m) \right\} /(t-t_m)^{n_m} \qquad (5.2.28)$$

may be employed. Generally, whether using symbolic or numerical means, programs are more comfortable calculating derivatives than integrals. Consequently, Eq.(5.2.28) will be preferred in general though there may be special circumstances where the integral formulation may be more efficient.

It is essential to compute values with a program which will cope with any desired accuracy. Usually this means working with something like MATHE-MATICA or MAPLE. With these there is a choice between finding results symbolically first or proceeding entirely numerically provided that sufficient figures are deployed to attain the required precision. The latter may well be much faster than the former but needs to be accompanied by an estimate of the reliability of the answers.

Example 5.2.1 The integral

$$\int_0^\infty e^{-xt} t^{-\frac{1}{2}} (1+t)^{-\frac{1}{2}} dt$$

does not appear to satisfy Eq.(5.2.1). However, by writing

$$(1+t)^{-\frac{1}{2}} = \frac{1+t}{(1+t)^{3/2}},$$

it can be split into two parts, with different values of μ, which do comply with Eq.(5.2.1) taking $p = 1$. Since μ does not play a crucial role in our estimates the technique described above can be applied to the integral.

When $x = 10$ the value of the integral is

0.54780756431351898687

correct to 20 decimal places. The successive approximations obtained by minimising the remainder at each stage are

	Correct$-$Approx	
0.54780433398410821970	$+3.230 \times 10^{-6}$	$n_0 = 10$
0.54780756444701134708	-1.335×10^{-10}	$n_1 = 13$
0.54780756431351898445	$+2.420 \times 10^{-18}$	$n_2 = 25$

The approximations were calculated by MATHEMATICA and are shown to 20 decimal places. The systematic improvement in the approximation is transparent but the third has involved an expansion to 48 terms.

Example 5.2.2 Here we consider

$$\int_0^\infty e^{-xt} t^{-1/6} (1+t)^{-1/6} dt$$

when $x = 16$. Its value, correct to 29 places of decimal, is

0.11107566172526420997574644654.

The successive approximations, with 27 figures retained, are

	Correct−Approx	
0.111075661089504605126050935	$+6.358 \times 10^{-10}$	$n_0 = 16$
0.111075661725264352424075563	-1.424×10^{-16}	$n_1 = 21$
0.111075661725264209975746447		$n_2 = 40$

The error in the third approximation is of the order 10^{-28} or better but requires 77 terms to achieve such high accuracy.

5.3   Extensions

The theory of the preceding section can be extended to other integrals. Consider

$$\int_0^\infty e^{-zt}t^\mu f(t)dt \tag{5.3.1}$$

where $f(t)$ satisfies Eq.(5.2.1) and z is complex. The bounds on R_m are unaffected provided that x is the real part of z. Therefore, so long as the real part of z is an unlimited positive number we can apply the previous method without any alteration to t_m or n_m.

Example 5.3.1 Take the integral of Eq.(5.3.1) with $\mu = -\frac{1}{2}$, $f(t) = (1+t)^{-\frac{1}{2}}$ and $z = 10 + 10i$. To 20 decimal places the value of the integral is

$$0.43207564347829846063 - 0.17324024391515304705i.$$

The approximations are

	Correct−Approx	
0.43207565255549401316	-9.077×10^{-9}	$n_0 = 10$
$-0.17324013824816036606i$	$-1.057 \times 10^{-7}i$	
0.43207564347838468606	-8.622×10^{-14}	$n_1 = 13$
$-0.17324024391514043347i$	$-1.261 \times 10^{-14}i$	
0.43207564347829846109	-4.662×10^{-19}	$n_2 = 25$
$-0.17324024391515304682i$	$-2.274 \times 10^{-19}i$	

Although z is complex no more terms are necessary to achieve comparable accuracy than when its imaginary part is zero.

A wider class of integrands can be encompassed by relaxing Eq.(5.2.1) to

$$\left|f^{(n)}(t)\right| \le \frac{(b+n)!K}{q^n(p+t)^{n+1}} \quad (t \ge 0) \tag{5.3.2}$$

with $0 < q \le 1$. The analysis of the preceding section can be repeated and the optimal remainder is attained by taking

$$t_m = p\{(q+1)^m - 1\}. \tag{5.3.3}$$

The demonstration that n_m increases with m may fail, however, if q is too small. If it is wished that the growth of n_m should be preserved it is best that q should be somewhat greater than $\frac{1}{2}$.

Analytic continuation of Eq.(5.3.1) is permissible when f is regular in a suitable region. The contour of integration is moved to some convenient ray from the origin and then this line is taken as the real axis. Thereafter, calculations can be carried out as if Eq.(5.3.1) were being dealt with; the added facility of Eq.(5.3.2) will be available also.

Example 5.3.2 Here the analytic continuation of

$$\int_0^\infty e^{-zt} t^{-\frac{1}{2}} (1+t)^{-\frac{1}{2}} dt$$

will be discussed. The case $z = 10i$ cannot be handled as in Example 5.3.1 because the real part of z is not sufficiently large and, indeed, the integral is not absolutely convergent. Nevertheless, the function can be continued analytically as in Section 2.2. By deforming the contour into the negative imaginary axis we find that the function is represented by

$$e^{-\pi i/4} \int_0^\infty e^{izt} t^{-\frac{1}{2}} (1-it)^{-\frac{1}{2}} dt \qquad (5.3.4)$$

for $0 < \mathrm{ph}\, z < \pi$. Here, after a similar modification to that in Example 5.2.1, Eq.(5.3.2) applies with $q = 1/\sqrt{2}$ and $p = 1$. Therefore, the method of Section 5.2 can be invoked with, of course, the change in Eq.(5.3.3).

With $z = 10i$ in Eq.(5.3.4) the value is

$$0.4049774294448407 - 0.3855795286055832i$$

to 16 places of decimals. The approximations are

	Correct−Approx	
0.4049795369094827	-2.107×10^{-6}	$n_0 = 7$
$-0.3855697462252094i$	$-9.782 \times 10^{-6}i$	
0.4049774237506989	$+5.694 \times 10^{-9}$	$n_1 = 7$
$-0.3855795493517057i$	$+2.075 \times 10^{-8}i$	
0.4049774294533430	-8.502×10^{-12}	$n_2 = 12$
$-0.3855795285972340i$	$-8.349 \times 10^{-12}i$	

It can be seen that having q less than 1 reduces the number of terms at each stage and causes some deterioration of accuracy.

As a final illustration of the wide applicability of the method the analytic continuation of the original function to $z = -10 + 15i$ is examined. This can be determined by substitution in Eq.(5.3.4). The value to aim for is

$$0.20294729303814455064 - 0.36829281936523898961i$$

to 20 places of decimal. The approximations are

	Correct−Approx	
0.20294729784081445049	-4.803×10^{-9}	$n_0 = 11$
$-0.368292812609221553526i$	$-6.756 \times 10^{-9}i$	
0.20294729303800567722	$+1.389 \times 10^{-13}$	$n_1 = 11$
$-0.36829281936492076857i$	$-3.182 \times 10^{-13}i$	
0.20294729303814454186	$+8.781 \times 10^{-18}$	$n_2 = 18$
$-0.36829281936523902034i$	$+3.073 \times 10^{-17}i$	

5.4   Stieltjes transforms

The Stieltjes transforms to be considered are of the form

$$F(z) = \int_0^\infty \frac{f(t)t^\mu}{t+z} e^{-t} dt \qquad (5.4.1)$$

which defines $F(z)$ for all complex z except those on the negative real axis. Interest centres on the behaviour of $F(z)$ when $|z|$ is unlimited. It will be assume that μ is limited and $\mu > -1$. Convergence of the integral is guaranteed if

$$|f(t)| \le Ke^{\sigma t} \qquad (5.4.2)$$

where $\sigma \le 1 - \delta_1$, δ_1 being standard and positive. If $z = \rho e^{i\theta}$ study of Eq.(5.4.1) is restricted to $|\theta| \le \pi - \delta$ where δ is standard and positive.

To a first approximation the t in the denominator can be ignored. This suggests introducing the expansion

$$\frac{1}{t+z} = \frac{1}{z}\sum_{m=0}^{n-1}\left(-\frac{t}{z}\right)^m + \frac{(-)^n t^n}{z^n(t+z)}.$$

Then

$$F(z) = \sum_{m=0}^{n-1}\frac{(-)^m}{z^{m+1}}\int_0^\infty f(t)t^{\mu+m}e^{-t}dt + R_0(n,z) \qquad (5.4.3)$$

where

$$R_0(n,z) = \frac{(-)^n}{z^n}\int_0^\infty \frac{f(t)t^{\mu+n}}{t+z}e^{-t}dt. \qquad (5.4.4)$$

The aim is to choose n to make the remainder suitably small. First, change the variable of integration by putting $t = \rho u$; then

$$R_0(n,z) = (-)^n\rho^\mu e^{-in\theta}\int_0^\infty \frac{f(\rho u)u^{\mu+n}}{u+e^{i\theta}}e^{-\rho u}du. \qquad (5.4.5)$$

Since $\left| u + e^{i\theta} \right| \geq \sin \delta$

$$|R_0(\nu, z)| \leq \frac{(\mu + \nu)! K}{\rho^\nu (1 - \sigma)^{\mu+\nu} \sin \delta} \tag{5.4.6}$$

by virtue of Eq.(5.4.2). For unlimited ν, Stirling's formula can be inserted in Eq.(5.4.6). Then the right-hand side is a minimum when $\nu = \rho(1 - \sigma)$ to within an infinitesimal. Therefore, in Eq.(5.4.3), take $n = n_0$ where

$$n_0 = [\rho(1 - \sigma)] \tag{5.4.7}$$

and the remainder is exponentially small.

To refine R_0 note that Eq.(5.4.6) stems from a neighbourhood of $u = u_1$ where

$$-\rho(1 - \sigma) + (\mu + \nu)/u_1 = 0.$$

In view of the choice of ν this means that

$$u_1 = 1 \tag{5.4.8}$$

if an infinitesimal is neglected. Consequently, the denominator in $R_0(n_0, z)$ is expanded about $u = u_1$ with the result

$$R_0(n_0, z) = (-)^{n_0} \rho^\mu e^{-in_0\theta}$$
$$\times \left\{ \sum_{m=0}^{n-1} \frac{(-)^m}{(u_1 + e^{i\theta})^{m+1}} \int_0^\infty f(\rho u) u^{\mu+n_0} (u - u_1)^m e^{-\rho u} du + R_1(n, z) \right\} \tag{5.4.9}$$

where

$$R_1(n, z) = \frac{(-)^n}{(u_1 + e^{i\theta})^n} \int_0^\infty \frac{f(\rho u) u^{\mu+n_0}}{u + e^{i\theta}} (u - u_1)^n e^{-\rho u} du. \tag{5.4.10}$$

Thus

$$|R_1(\nu, z)| \leq \frac{K}{|u_1 + e^{i\theta}|^\nu \sin \delta} \int_0^\infty u^{\mu+n_0} |u - u_1|^\nu e^{-\rho(1-\sigma)u} du.$$

The integrand is stationary at $u = u_2$ where

$$\frac{\mu + n_0}{u_2} + \frac{\nu}{u_2 - u_1} - \rho(1 - \sigma) = 0 \tag{5.4.11}$$

so long as $u_2 > u_1$ as will be checked shortly. Then the exponential behaviour of $|R_1(\nu, z)|$ is essentially

$$(\mu + n_0) \ln u_2 + \nu \ln(u_2 - u_1) - \rho(1 - \sigma)u_2 - \nu \ln \left| u_1 + e^{i\theta} \right|$$

which is stationary for

$$u_2 - u_1 = \left| u_1 + e^{i\theta} \right|. \tag{5.4.12}$$

With u_2 given by Eq.(5.4.12) the number of terms in Eq.(5.4.9) is fixed by $n = n_1$ where $n_1 = [\nu]$, ν being determined by Eq.(5.4.11).

Now the denominator in $R_1(n_1, z)$ is expanded about $u = u_2$ to n_2 terms where $n_2 = [\nu]$ with

$$\frac{\mu + n_0}{u_3} + \frac{n_1}{u_3 - u_1} + \frac{\nu}{u_3 - u_2} - \rho(1 - \sigma) = 0$$

and $u_3 - u_2 = \left| u_2 + e^{i\theta} \right|$.

Clearly, the process can be continued. In general,

$$u_{s+1} - u_s = \left| u_s + e^{i\theta} \right| \tag{5.4.13}$$

and $n_s = [\nu]$ where

$$\frac{\mu + n_0}{u_{s+1}} + \sum_{r=1}^{s-1} \frac{n_r}{u_{s+1} - u_r} + \frac{\nu}{u_{s+1} - u_s} - \rho(1 - \sigma) = 0. \tag{5.4.14}$$

There is an alternative to Eq.(5.4.13) for determining u_s. From Eq.(5.4.13)

$$\begin{aligned}
u_{s+1} + e^{i\theta} &= u_s + e^{i\theta} + \left| u_s + e^{i\theta} \right| \\
&= 2 \left| u_s + e^{i\theta} \right| e^{i\phi_s/2} \cos(\phi_s/2)
\end{aligned}$$

where ϕ_s is the phase of $u_s + e^{i\theta}$. Evidently $\phi_{s+1} = \phi_s/2$. Since it is clear that $u_1 + e^{i\theta} = 2e^{i\theta/2} \cos(\theta/2)$, $\phi_1 = \theta/2$ and so $\phi_s = \theta/2^s$. Now Eq.(5.4.13) implies that

$$\begin{aligned}
u_{s+1} - u_s &= 2 \left| u_{s-1} + e^{i\theta} \right| \cos(\phi_{s-1}/2) \\
&= 2^s \cos\frac{\theta}{2^s} \cos\frac{\theta}{2^{s-1}} \cdots \cos\frac{\theta}{2}
\end{aligned} \tag{5.4.15}$$

on repeated application. Thus, with $u_1 = 1$, $u_2, u_3, \ldots$ can be found from Eq.(5.4.15). Naturally, for numerical purposes, often it will be more efficient to employ Eq.(5.4.13).

Remark The condition of Eq.(5.4.2) forces $f(t)$ to be bounded at the origin. Such a restriction is unnecessary. It is sufficient for $f(t)t^\mu$ to be integrable in a neighbourhood $(0, \epsilon)$ of the origin and to satisfy Eq.(5.4.2) outside. The change affects Eq.(5.4.6) only infinitesimally so that the succeeding argument is unaltered.

Example 5.4.1 Choose $f(t) = K_0(t)$ where K_0 is the modified Bessel function and $\mu = -\frac{1}{2}$. Then it can be shown that

$$F(z) = \pi e^z K_0(z)/z^{\frac{1}{2}}.$$

According to the remark above the preceding theory can be applied with $\sigma = -1$ since $K_0(t)$ decays exponentially for large enough t.

Now

$$F(5i) = 0.0192710158058762 - 0.7853857878891211i$$

to 16 places of decimal and the approximations are

	Correct$-$Approx	
0.0192752428015537	-4.227×10^{-6}	$n_0 = 10$
$-0.7853904081417598i$	$+4.620 \times 10^{-6}i$	
0.0192710242449071	-8.439×10^{-9}	$n_1 = 8$
$-0.7853857917780257i$	$+3.889 \times 10^{-9}i$	
0.0192710158078479	-1.972×10^{-12}	$n_2 = 16$
$-0.7853857878949455i$	$+5.824 \times 10^{-12}i$	

Notice that, in contrast to the Laplace integral, the value of n_m does not necessarily increase with m.

As z approaches the negative real axis the integral tends to become singular and the approximations are likely to become less accurate. An indication of this happening can be seen by taking $z = 5(i-1)/\sqrt{2}$. Then

$$F(z) = -0.5547417804 - 0.5779236647i$$

and the approximations are

	Correct$-$Approx	
-0.5547518491	0.0000101	$n_0 = 10$
$-0.5779268407i$	$+3.176 \times 10^{-6}i$	
-0.5547419221	1.417×10^{-7}	$n_1 = 3$
$-0.5779242095i$	$+5.448 \times 10^{-7}i$	
-0.5547419520	1.716×10^{-7}	$n_2 = 6$
$-0.5779233259i$	$-3.388 \times 10^{-7}i$	

5.5 Stokes' phenomenon

The expansion of $F(z)$ in Section 5.4 was limited to the sector $|\mathrm{ph}\, z| < \pi$. One reason is that, if the contour of integration cannot be moved from the real axis, the integrand of the Stieltjes transform becomes singular when z is on the negative real axis. However, if f is suitably regular in a sector including the positive real axis it is possible to shift the contour and so supply an analytic continuation of $F(z)$ across the negative real axis. Of course, the deformation effectively captures a pole when z is near the negative real axis (compare Section

4.2) and this extra contribution, although exponentially small, can be of the same order of magnitude as the remainders considered in Section 5.4. Moreover, as the phase of z increases from π (or decreases from $-\pi$), the exponential decay becomes less pronounced and the extra term may well have a dominating influence on the asymptotic expansion of $F(z)$. In other words, the negative real axis will be a Stokes line. The result is that a more complicated analysis than that of Section 5.4 is necessary to accommodate the asymptotic behaviour of $F(z)$ for z in a neighbourhood of the negative real axis.

The starting point is the expansion of Eq.(5.4.3), namely

$$F(z) = \sum_{m=0}^{n-1} \frac{(-)^m}{z^{m+1}} \int_0^\infty f(t) t^{\mu+m} e^{-t} dt + R_0(n, z) \qquad (5.5.1)$$

where

$$R_0(n, z) = \frac{(-)^n}{z^n} \int_0^\infty \frac{f(t) t^{\mu+n}}{t+z} e^{-t} dt. \qquad (5.5.2)$$

The extra term referred to above is contained in R_0 and, to take care of it, we shall make certain assumptions about $f(t)$. It will be supposed that, for $\alpha_2 \geq \text{ph}\, t \geq \alpha_1$ where $\pi/2 - \delta \geq \alpha_2 > 0 > \alpha_1 \geq -\pi/2 + \delta$ (δ being a small standard positive number),

$$f(t) = f_0(t) e^{\sigma t} \qquad (5.5.3)$$

with $\sigma < 1$. The function f_0 is regular in the specified sector except possibly at the origin and satisfies

$$\left| f_0^{(m)}(t) \right| \leq \frac{(b_1 + m)!}{(\mathcal{R}t)^{\alpha+m}} K_1 + \frac{(b_2 + m)!}{(\mathcal{R}t)^{\beta+m}} K_2 \qquad (5.5.4)$$

where $0 < \beta \leq \alpha < 1$. The term in K_1 is to allow for a possible singularity at the origin whereas that in K_2 governs what happens at infinity. It will be seen later that more freedom can be permitted to β than has been prescribed so far.

Now rewrite Eq.(5.5.2) as

$$\begin{aligned} R_0(n, z) &= \frac{(-)^n}{z^n} \int_0^\infty \frac{f_0(t) - f_0(-z)}{t+z} t^{\mu+n} e^{-(1-\sigma)t} dt \\ &+ \frac{(-)^n}{z^n} f_0(-z) \int_0^\infty \frac{t^{\mu+n}}{t+z} e^{-(1-\sigma)t} dt. \end{aligned} \qquad (5.5.5)$$

The second integral can be expressed in terms of the function $J_0(\mu, z)$ discussed in Section 4.2 and defined by

$$J_0(\mu, z) = \int_0^\infty \frac{t^\mu e^{-t}}{t+z} dt.$$

As for z it will be sufficient to examine what occurs when ph z passes through π since the case when ph z is near $-\pi$ can be dealt with in a similar manner. Therefore, take $z = \rho e^{i(\pi - \phi)}$ where $\alpha_2 \geq -\phi \geq \alpha_1$. Then

$$R_0(n, z) = \frac{(-)^n f_0(\rho e^{-i\phi})}{z^n (1 - \sigma)^{\mu + n}} J_0\{\mu + n, (1 - \sigma)z\} + r_0(n, z) \tag{5.5.6}$$

where

$$r_0(n, z) = \rho^\mu e^{in\phi} \int_0^\infty g(t) t^{\mu + n} e^{-\rho(1 - \sigma)t} dt \tag{5.5.7}$$

and

$$g(t) = \frac{f_0(\rho t) - f_0(\rho e^{-i\phi})}{t - e^{-i\phi}}. \tag{5.5.8}$$

Since $g(t)$ is regular at $t = e^{-i\phi}$ the formula of Eq.(5.5.6) supplies an analytic continuation of $R_0(n, z)$ covering the sector $\pi + \alpha_2 \geq \text{ph } z \geq \pi + \alpha_1$.

In order to have an estimate of the magnitude of r_0 it is of assistance to bound a certain integral. If $c > 0$, $a > 0$, $n > 0$, $m \geq 0$, $0 < \gamma < 1$ the substitution

$$u = \frac{c(1 - v)}{c(1 - v) + av}$$

gives

$$\int_0^1 \frac{(1 - u)^{n-1} u^m du}{\{c(1 - u) + au\}^{\gamma + m + n}}$$
$$= \frac{1}{c^{\gamma + n - 1} a^{\gamma + m}} \int_0^1 v^{n-1}(1 - v)^m \{c(1 - v) + av\}^{\gamma - 1} dv$$
$$\leq \frac{(n - 1)!(m + \gamma - 1)!}{(m + n + \gamma - 1)! c^n a^{\gamma + m}} \tag{5.5.9}$$

on suppressing the term av and employing the formula for the Beta function.

Now, from Eq.(5.5.8),

$$g(t) = \rho \int_0^1 f_0'[\rho\{(1 - u)e^{-i\phi} + ut\}] du.$$

Hence, from Eq.(5.5.4),

$$\left| g^{(m)}(t) \right| \leq \int_0^1 \left[\frac{(b_1 + m + 1)! K_1}{\rho^\alpha \{(1 - u)\cos\phi + ut\}^{\alpha + m + 1}} \right.$$
$$\left. + \frac{(b_2 + m + 1)! K_2}{\rho^\beta \{(1 - u)\cos\phi + ut\}^{\beta + m + 1}} \right] u^m du$$

for t real and positive. An invocation of Eq.(5.5.9) with $n = 1$, $c = \cos\phi$, $a = t$ furnishes

$$\left| g^{(m)}(t) \right| \leq \frac{(b_1 + m + 1)! K_1}{\rho^\alpha t^{\alpha + m}(m + \alpha)\cos\phi} + \frac{(b_2 + m + 1)! K_2}{\rho^\beta t^{\beta + m}(m + \beta)\cos\phi} \tag{5.5.10}$$

which bears a distinct similarity to Eq.(5.5.4) with t real.

A bound for $|r_0(\nu, z)|$ can be determined now from Eq.(5.5.7) and Eq.(5.5.10). Evidently, the main contribution to the integral with ν and ρ unlimited will come from a neighbourhood of $t = t_1$ where

$$\nu/t_1 = (1 - \sigma)\rho$$

to within an infinitesimal. The major factor in $|r_0(\nu, z)|$ is then

$$\exp\{\nu \ln t_1 - (1 - \sigma)\rho t_1\}$$

which is stationary with respect to ν when $t_1 = 1$. Thus, on taking

$$n = n_0 = [(1 - \sigma)\rho] \tag{5.5.11}$$

in Eq.(5.5.6), the remainder r_0 is exponentially damped. The number of terms in the expansion of Eq.(5.5.1) has been fixed now by n_0.

To proceed further with hyperasymptotics note that the performance of $r_0(n_0, z)$ is dictated by the integrand near $t = t_1$. Therefore, expand $g(t)$ about this point to obtain

$$r_0(n_0, z) = \rho^\mu e^{in_0\phi} \sum_{m=0}^{n-1} \frac{g^{(m)}(t_1)}{m!} \int_0^\infty t^{\mu+n_0}(t - t_1)^m e^{-(1-\sigma)\rho t}dt + r_1(n, z) \tag{5.5.12}$$

where

$$r_1(n, z) = \rho^\mu e^{in_0\phi} \int_0^\infty t^{\mu+n_0}(t - t_1)^n g_1(t)e^{-(1-\sigma)\rho t}dt \tag{5.5.13}$$

with

$$
\begin{aligned}
g_1(t) &= \left\{ g(t) - \sum_{m=0}^{n-1}(t - t_1)^m g^{(m)}(t_1)/m! \right\} \Big/ (t - t_1)^n \\
&= \frac{1}{(n-1)!} \int_0^1 (1 - u)^{n-1} g^{(n)}\{t_1 + u(t - t_1)\}du. \tag{5.5.14}
\end{aligned}
$$

By means of Eq.(5.5.10) and Eq.(5.5.9) we deduce that

$$\left| g_1^{(m)}(t) \right| \leq \frac{(b_1 + m + n + 1)!(m + \alpha)!K_1}{(m + n + \alpha)!\rho^\alpha t_1^n t^{\alpha+m} \cos\phi} + \frac{(b_2 + m + n + 1)!(m + \beta)!K_2}{(m + n + \beta)!\rho^\beta t_1^n t^{\beta+m} \cos\phi}. \tag{5.5.15}$$

In view of Eq.(5.5.15) the major contribution in bounding $r_1(\nu, z)$ by the formula of Eq.(5.5.13) comes from near $t = t_2$ where

$$\frac{n_0}{t_2} + \frac{\nu}{t_2 - t_1} = (1 - \sigma)\rho \tag{5.5.16}$$

and the result is stationary with respect to ν if

$$t_2 = 2t_1.$$

Hence, on putting $n = n_2 = [\nu]$ with ν specified by Eq.(5.5.16), the remainder $r_1(n_2, z)$ is exponentially small.

It is transparent that the pattern is similar to that of Section 5.2. The principal differences are that $t_m = 2^{m-1}$ and n_m is given by an obvious generalisation of Eq.(5.5.16). Consequently, the analysis of successive remainders will not be pursued further.

Note It is not necessary to have precise values for α and β in a specific calculation; it suffices to know that they exist. For, they are used only to verify theoretically that the remainders are exponentially damped; they do not figure in the definitions of t_m or n_m which are the sole quantities required to specify which terms are evaluated in the expansion. Furthermore, the lower bound on β can be dropped so long as β remains standard. If $\beta \leq 0$ multiplication of the numerator and denominator of the term involving K_2 by t^τ where $0 < \tau$ and $0 < \tau + \beta \leq \alpha$ will alter μ but allow the preceding argument to go through unchanged.

Example 5.5.1 In Example 5.4.1 the case $f(t) = K_0(t)$, $\mu = -\frac{1}{2}$ and $z = 5(i-1)/\sqrt{2}$ was evaluated. Here the calculation is repeated using the method of this section. For comparison

$$F(z) = -0.554741780397399 - 0.577923664708075i.$$

The approximations are

	Correct−Approx	
−0.554751849096424	$+1.007 \times 10^{-5}$	$n_0 = 10$
−0.577926840696955i	$+3.176 \times 10^{-6}i$	
−0.554744159410270	$+2.379 \times 10^{-6}$	J_0
−0.577919955823531i	$-3.709 \times 10^{-6}i$	
−0.554741764899744	-1.550×10^{-8}	$n_1 = 5$
−0.577923682973878i	$+1.827 \times 10^{-8}i$	
−0.554741780659114	$+2.617 \times 10^{-10}$	$n_2 = 11$
−0.577923664424185i	$-2.839 \times 10^{-10}i$	

The second row shows the effect of adding in the J_0 term of Eq.(5.5.6). It has a relatively minor effect but the subsequent approximations are more accurate than in Example 5.4.1. Thus the method of this section offers a useful improvement but at the price of evaluating an additional 10 terms.

Example 5.5.2 With the same f and μ as in Example 5.4.1 put $z = 5e^{\pi i}$. Then

$$F(z) = -0.810116342676668349 - 0.000034942025501635i.$$

The corresponding approximations are

	Correct$-$Approx	
-0.810110594286227767	-5.748×10^{-6}	$n_0 = 10$
	$-3.494 \times 10^{-5}i$	
-0.810112046671819862	-4.296×10^{-6}	J_0
$-0.000034942025501635i$		
-0.810116366083665133	$+2.341 \times 10^{-8}$	$n_1 = 5$
$-0.000034942025501635i$		
-0.81011634229593623	-3.807×10^{-10}	$n_2 = 11$
$-0.000034942025501635i$		

It can be seen that the principal purpose of the J_0 term in this example is to provide the imaginary part of the approximation. This it does to extremely high accuracy so that the error estimates of the imaginary part have been omitted after its inclusion.

Exercises on Chapter 5

1. Repeat Example 5.2.1 with $x = 5$.
2. Repeat Example 5.2.2 with $x = 10$.
3. Repeat Example 5.3.1 with $z = 4 + 3i$.
4. By analytic continuation determine the integral in Example 5.2.2 when $z = 10i$.
5. Repeat Example 5.4.1 for $z = 10$ and $z = 10i$.
6. The formula

$$\pi e^z K_\nu(z) = z^{\frac{1}{2}} \cos \nu\pi \int_0^\infty \frac{K_\nu(t)t^{-\frac{1}{2}}}{t+z} e^{-t} dt$$

is valid for $|\mathcal{R}\nu| < \frac{1}{2}$. Check the method of Section 5.4 for $z = 10$ and $z = 10i$ when $\nu = 1/4$ and $\nu = i/4$.

7. Explain how to continue analytically the formula of Exercise 6 across the negative real axis and check it with $z = 10e^{\pi i}$.

8. Repeat Example 5.5.2 with $z = 5e^{-\pi i}$.

9. The analysis of Section 5.5 requires $\mathcal{R}(z) < 0$. Show, by computing the integral of Example 5.5.1 with $z = 5i$, that the method of Section 5.5 supplies a satisfactory answer nevertheless. Is it possible to modify the analysis to include the imaginary axis?

Chapter 6
DIFFERENTIAL EQUATIONS

6.1 The WKB approximation

A customary form for a linear differential equation of the second order is

$$\frac{d^2y}{dz^2} + p(z)\frac{dy}{dz} + q(z)y = 0. \tag{6.1.1}$$

The transformation

$$y = w \exp\{-\tfrac{1}{2}\int^z p(t)dt\}$$

converts Eq.(6.1.1) to

$$\frac{d^2w}{dz^2} - g(z)w = 0 \tag{6.1.2}$$

where

$$g(z) = \tfrac{1}{2}p'(z) + \tfrac{1}{4}p^2(z) - q(z). \tag{6.1.3}$$

Usually the version in Eq.(6.1.2) will be concentrated on in this chapter.

To look for approximate solutions of Eq.(6.1.2) we attempt to transform it into the corresponding differential equation for the exponential. Put

$$w = W/\{g(z)\}^{1/4}. \tag{6.1.4}$$

Then Eq.(6.1.2) becomes

$$\frac{d^2W}{dz^2} - \frac{g'}{2g}\frac{dW}{dz} + \left\{g^{1/4}\frac{d^2}{dz^2}\left(\frac{1}{g^{1/4}}\right) - g\right\}W = 0$$

which, after the change of variable

$$\xi(z) = \int^z g^{\frac{1}{2}}(t)dt, \tag{6.1.5}$$

goes over to

$$\frac{d^2W}{d\xi^2} - (1+\psi)W = 0 \tag{6.1.6}$$

where

$$\psi(\xi) = -\frac{1}{g^{3/4}}\frac{d^2}{dz^2}\left(\frac{1}{g^{1/4}}\right). \qquad (6.1.7)$$

If now ψ is negligible compared with unity over some range of ξ it is evident from Eq.(6.1.6) that W is effectively $e^{\pm\xi}$. Therefore, for negligible ψ, Eq.(6.1.4) gives

$$w = (Ae^{\xi} + Be^{-\xi})/g^{1/4} \qquad (6.1.8)$$

as an approximate solution of Eq.(6.1.2); A and B are arbitrary constants whereas ξ is obtained from Eq.(6.1.5). The formula of Eq.(6.1.8) is known as the *WKB approximation* (sometimes the letters are in a different order and sometimes other initials may be quoted).

The validity or otherwise of Eq.(6.1.8) rests on ψ being sufficiently small. Clearly, trouble can be expected if g is small and, in particular, if the region under consideration contains a zero of $g(z)$. For the moment it will be supposed that such awkward features are excluded.

6.2   An error bound for the WKB approximation

The derivation of the approximation in the preceding section rests on the assumption that ψ is negligible but gives no indication of how much the approximate solution differs from the exact on account of the presence of ψ. In this section a bound for the error is determined.

The differential equation of Eq.(6.1.6) is exact. If the term ψW is moved to the right-hand side the general solution can be expressed as

$$W(\xi) = Ae^{\xi} + Be^{-\xi} + \frac{1}{2}\int_{\alpha_1}^{\xi}(e^{\xi-v} - e^{v-\xi})\psi(v)W(v)dv.$$

To examine how a solution deviates from e^{ξ} put

$$W(\xi) = e^{\xi}\{1 + e(\xi)\}$$

where $e(\xi)$ represents the error due to ψ. Then

$$1 + e(\xi) = A + Be^{-2\xi} + \frac{1}{2}\int_{\alpha_1}^{\xi}(1 - e^{2v-2\xi})\psi(v)\{1 + e(v)\}dv.$$

The conditions $e(\xi) \to 0$, $e'(\xi) \to 0$ as $\xi \to \alpha_1$ ensure that $W(\xi)$ is close to e^{ξ} as $\xi \to \alpha_1$; they enforce $A = 1$, $B = 0$. Consequently

$$e(\xi) = \frac{1}{2}\int_{\alpha_1}^{\xi}(1 - e^{2v-2\xi})\psi(v)\{1 + e(v)\}dv \qquad (6.2.1)$$

provides a solution of Eq.(6.1.6) which is near to e^ξ when ξ is in the neighbourhood of α_1.

Solving the integral equation in Eq.(6.2.1) is accomplished by iteration in which successive estimates are related by

$$e_n(\xi) = \tfrac{1}{2}\int_{\alpha_1}^{\xi}(1 - e^{2v-2\xi})\psi(v)\{1 + e_{n-1}(v)\}dv. \qquad (6.2.2)$$

If e_n tends to a suitable limit as $n \to \infty$ then the scheme of Eq.(6.2.2) will solve Eq.(6.2.1). The iteration is started with $e_0(\xi) \equiv 0$ so that

$$e_1(\xi) = \tfrac{1}{2}\int_{\alpha_1}^{\xi}(1 - e^{2v-2\xi})\psi(v)dv. \qquad (6.2.3)$$

Then e_n can be estimated via

$$e_n(\xi) = \sum_{m=1}^{n}\{e_m(\xi) - e_{m-1}(\xi)\} \qquad (6.2.4)$$

where

$$e_m(\xi) - e_{m-1}(\xi) = \tfrac{1}{2}\int_{\alpha_1}^{\xi}(1 - e^{2v-2\xi})\psi(v)\{e_{m-1}(v) - e_{m-2}(v)\}dv \qquad (6.2.5)$$

from Eq.(6.2.2).

Bounding the integrals in Eqs.(6.2.3) and (6.2.5) is not quite straightforward because α_1, ξ and ψ may be complex in general. Identify points on the path of integration by the arc length s measured from α_1. Let $s = s_0$ when $v = \xi$. Then the integral in Eq.(6.2.3) can be expressed as

$$\int_0^{s_0}(1 - e^{2v-2\xi})\psi(v)\frac{dv}{ds}ds.$$

If $\mathcal{R}(v)$ does not decrease on the contour $\mathcal{R}(v - \xi) \le 0$ and $\left|1 - e^{2v-2\xi}\right| \le 2$. Accordingly

$$|e_1(\xi)| \le \int_0^{s_0}|\psi(v)|\,ds. \qquad (6.2.6)$$

Now suppose that, for some m,

$$|e_m(\xi) - e_{m-1}(\xi)| \le \left\{\int_0^{s_0}|\psi(v)|\,ds\right\}^m \Big/ m! \qquad (6.2.7)$$

which is true for $m = 1$ by virtue of Eq.(6.2.6). From Eq.(6.2.5)

$$\begin{aligned}
|e_{m+1}(\xi) - e_m(\xi)| &\le \int_0^{s_0}|\psi(v)|\left\{\int_0^s|\psi(v)|\,ds\right\}^m ds\Big/ m! \\
&\le \left\{\int_0^{s_0}|\psi(v)|\,ds\right\}^{m+1}\Big/ (m+1)!
\end{aligned}$$

and induction verifies Eq.(6.2.7) for all m. It follows from Eq.(6.2.4) that

$$|e_n(\xi)| \leq \sum_{m=1}^{n} \left\{ \int_0^{s_0} |\psi(v)| \, ds \right\}^m / m!$$

$$\leq \exp\left\{ \int_0^{s_0} |\psi(v)| \, ds \right\} - 1. \qquad (6.2.8)$$

Now if, for some positive s', $\int_0^{s'} |\psi(v)| \, ds$ is finite the integral in Eq.(6.2.8) is bounded for $0 \leq s_0 \leq s'$. In particular, if s' corresponds to the point α_2, the solution of Eq.(6.2.1) has been found on the path joining α_1 to α_2 and an explicit bound for it at any point of the path has been obtained. Since

$$e_n'(\xi) = \int_{\alpha_1}^{\xi} e^{2v - 2\xi} \psi(v) \{ 1 + e_{n-1}(v) \} dv,$$

$$e_m'(\xi) - e_{m-1}'(\xi) = \int_{\alpha_1}^{\xi} e^{2v - 2\xi} \psi(v) \{ e_{m-1}(v) - e_{m-2}(v) \} dv.$$

Invoking Eq.(6.2.7) we deduce that $|e_n'(\xi)|$ satisfies the same bound as $|e_n(\xi)|$.

These results may be combined as

Theorem 6.2.1 *Let v be a typical point on the path joining α_1 and ξ where the arc-length from α_1 is s. If $s = s_0$ when $v = \xi$ and v does not decrease on the path then there is an exact solution $e^{\xi}\{1 + e(\xi)\}$ of Eq.(6.1.6) such that $|e(\xi)|$ and $|e'(\xi)|$ are both bounded above by*

$$\exp\left\{ \int_0^{s_0} |\psi(v)| \, ds \right\} - 1.$$

The corresponding theorem for a solution like $e^{-\xi}$ can be achieved by changing the sign of ξ. In this case the solution must be started at the other end, say the point α_2 mentioned above.

Theorem 6.2.2 *If, in Theorem 6.2.1, α_1 is replaced by α_2 and v does not increase on the path there is an exact solution $e^{-\xi}\{1 + e(\xi)\}$ where $e(\xi)$ and $e'(\xi)$ satisfy the same bounds as in Theorem 6.2.1.*

Several points about these theorems should be noted. They provide only bounds to the errors not sharp estimates and the bounds may differ from the true errors by a substantial margin. As an indication let ξ be real and the path of integration the real axis. Then, the inequality $\left| 1 - e^{2v - 2\xi} \right| \leq 1$ would be valid just before Eq.(6.2.6) with the effect of halving the argument of the exponential in Eq.(6.2.8). Thereby the bound could be reduced significantly though perhaps still some way from the true error.

There is no necessity for $\psi(v)$ to be bounded at α_1 (or α_2). So long as its modulus is integrable with respect to arc-length the bound will hold.

Finding paths of integration which have the requisite properties may not be easy. Unfortunately, there seem to be no general rules to offer clues on how to make a choice and each differential equation has to be treated on its own merits.

When $\mathcal{R}(\xi)$ augments in going from α_1 to α_2 the solution W_1 ($\sim e^{\xi}$) is much larger than W_2 ($\sim e^{-\xi}$) near α_2. W_1 is said to be *dominant* at α_2 whereas W_2 is called *recessive* at α_2. Obviously the roles are reversed at α_1. If we know that a solution behaves like $e^{-\xi}$ near α_2 than it can be no other than W_2 (to within a constant multiple). However, the behaviour of e^{ξ} near α_2 does not identify W_1 uniquely because $W_1 + W_2$ exhibits the same behaviour. On the other hand, when ξ is purely imaginary on the path of integration, both W_1 and W_2 are of the same order of magnitude; neither dominates the other. Yet each can be identified uniquely by prescribing the appropriate behaviour near α_2 or α_1.

Example 6.2.1 Consider solutions of the differential equation

$$\frac{d^2w}{dx^2} = (x^2 + 1)w$$

for real x as $x \to \infty$. An attempt to apply the preceding theory immediately by taking $\xi = x$ and $\psi(\xi) = \xi^2$ would fail because the error terms grow without bound as $x \to \infty$. Therefore, it is necessary to return to the general theory of Section 6.1 and put

$$\xi = \int^x (1 + t^2)^{\frac{1}{2}} dt.$$

For large x this gives

$$\xi = \tfrac{1}{2}x^2 + \tfrac{1}{2}\ln x + \text{constant}$$

apart from terms which tend to zero as $x \to \infty$. It follows from Eq.(6.1.7) that

$$|\psi(\xi)| \sim 1/\xi^2$$

for large x (or ξ). Since ψ is integrable as $\xi \to \infty$ the error terms in Theorems 6.2.1 and 6.2.2 are bounded. Hence there are solutions which are constant multiples of $\exp(\tfrac{1}{2}x^2)$ and $\exp(-\tfrac{1}{2}x^2)/x$ as $x \to \infty$.

6.3   Effect of a parameter

Quite often applications entail solving

$$\frac{d^2w}{dz^2} = \{k^2 g(z) + h(z)\}w \tag{6.3.1}$$

where k is a parameter in which $|k|$ is large. The differential equation in Eq.(6.3.1) is tackled by the same method as that of Section 6.1 i.e. one puts $w = W/\{g(z)\}^{1/4}$ and

$$\xi(z) = k \int^z g^{\frac{1}{2}}(t) dt.$$

Then

$$\frac{d^2W}{d\xi^2} = (1 + \psi)W \tag{6.3.2}$$

where

$$\psi(\xi) = \frac{1}{k^2}\left\{\frac{h}{g} - \frac{1}{g^{3/4}}\frac{d^2}{dz^2}\left(\frac{1}{g^{1/4}}\right)\right\}. \tag{6.3.3}$$

Now we are in a position to apply Theorems 6.2.1 and 6.2.2. Observe that the parameter is not involved in evaluating the integral of the error bound. Therefore the error bound will be uniform with respect to z for the path from α_1 to α_2 and will be smaller the larger $|k|$ is.

Example 6.3.1 The differential equation

$$\frac{d^2w}{dx^2} = (x^2 + k^2)w,$$

with k real, cannot be handled directly as has been seen in Example 6.2.1. However, the substitution $x = ky$ transforms it to

$$\frac{d^2w}{dy^2} = k^4(y^2 + 1)w$$

which is of the form of Eq.(6.3.1) with h absent. Consequently, put

$$\xi = k^2 \int^y (t^2 + 1)^{\frac{1}{2}}dt.$$

Apart from the presence of the parameter the estimates of Example 6.2.1 can be employed. There are solutions which are constant multiples of $\exp(\frac{1}{2}k^2y^2)y^{\frac{1}{2}k^2-\frac{1}{2}}$ and $\exp(-\frac{1}{2}k^2y^2)y^{-\frac{1}{2}k^2-\frac{1}{2}}$. The error in the recessive solution is $O(1/k^2y^2)$ from Theorem 6.2.2. The conversion back to x is immediate.

Example 6.3.2 Here is considered

$$\frac{d^2w}{dz^2} = \left(4z^2 - \frac{1}{4z^2}\right)w.$$

Taking $g(z) = 4z^2$, $h(z) = -1/4z^2$ we have

$$\xi = z^2, \quad \psi = -1/4z^4 = -1/4\xi^2.$$

Thus there is a solution

$$\frac{1}{z^{\frac{1}{2}}}\exp(-z^2)\{1 + e(z)\}$$

which is recessive as $z \to \infty$.

It is of interest to see how the bound for the error varies with the path chosen from z to the point at infinity. Pick α_2 to be the point at infinity on the positive real axis. Let the point $\xi = re^{i\theta}$ correspond to z. A permissible path to α_2 in the ξ-plane when $0 \le \theta \le \frac{1}{2}\pi$ is shown in Fig. 6.3.1 since $\mathcal{R}(\xi)$ increases on it. Then

$$\int |\psi(v)| \, ds = (\theta + 1)/4r \qquad (6.3.4)$$

and $|e(z)|$ is bounded by $\exp\{(\theta + 1)/4 \, |z|^2\} - 1$ on the path in the z-plane obtained from Fig. 6.3.1 by the mapping $\xi = z^2$.

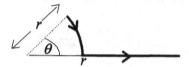

Figure 6.3.1 One path for the error bound

Other paths are possible. For example, see Fig. 6.3.2. On this path

$$\int |\psi(v)| \, ds = \frac{1}{4}\left(\frac{1}{r} - \frac{1}{\rho}\right) + \frac{\theta + 1}{4\rho}$$

(cf. Eq.(6.3.4)). Let $\rho \to \infty$; the terms in ρ disappear and

$$|e(z)| \le \exp(1/4 \, |z|^2) - 1.$$

This is a smaller bound than that derived for Fig. 6.3.1 and so offers a preferable path. The bound is applicable for $0 \le \mathrm{ph}\, z \le \pi/4$.

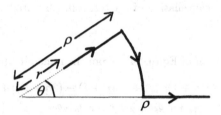

Figure 6.3.2 Another possible path

If $\mathcal{R}(\xi) < 0$ a path analogous to Fig. 6.3.2 is furnished by drawing a line vertically upwards from ξ until it meets the circle of radius ρ and then following the circumference of the circle as in Fig. 6.3.2. The details are left as an exercise.

6.4   Solutions in series

The WKB approximation gives a good idea of asymptotic performance. With more detailed information on $g(z)$ it is possible to extend the WKB approximation to an asymptotic expansion. The differential equation to be discussed is

$$\frac{d^2 w}{dz^2} = g(z) w \tag{6.4.1}$$

where $g(z)$ can be expanded in a convergent power series of the form

$$g(z) = \sum_{m=0}^{\infty} g_m / z^m \tag{6.4.2}$$

for $|z| > R$.

For large $|z|$

$$g^{\frac{1}{2}}(z) \sim g_0^{\frac{1}{2}} + g_1 / 2 g_0^{\frac{1}{2}} z$$

so that the WKB approximation contains an exponential factor

$$\exp\left\{ g_0^{\frac{1}{2}} z + \left(g_1 / 2 g_0^{\frac{1}{2}} \right) \ln z \right\}.$$

This suggests that a solution of Eq.(6.4.1) should be sought in the form

$$w = e^{\lambda z} z^{\mu} \sum_{m=0}^{\infty} a_m / z^m \tag{6.4.3}$$

with $a_0 \neq 0$.

If Eq.(6.4.3) is substituted in Eq.(6.4.1) and the factor $e^{\lambda z} z^{\mu}$ eliminated the constant terms cancel if

$$\lambda^2 = g_0 \tag{6.4.4}$$

since $a_0 \neq 0$. The cancellation of the terms in $1/z$ entails

$$2\lambda\mu = g_1 \tag{6.4.5}$$

with implementation of Eq.(6.4.4). From the remaining powers we have

$$-2\lambda(n+1)a_{n+1} + \{n(n+1) - (2n+1)\mu + \mu^2 - g_2\}a_n$$
$$= g_3 a_{n-1} + g_4 a_{n-2} + \cdots + g_{n+2} a_0 \tag{6.4.6}$$

for $n \geq 0$, the right-hand side being zero when $n = 0$. Evidently Eq.(6.4.6) permits the recursive determination of a coefficient a_{n+1} when its predecessors are known provided that $\lambda \neq 0$. Hence special consideration is required when either g_0 or g_1 is zero. The various possibilities when g_0 is zero and non-zero are now discussed separately.

(i) $g_0 = 0$, $g_1 \neq 0$

In this case return to Eq.(6.4.1) and make the change of variable $z = t^2$ followed by the substitution $w = t^{\frac{1}{2}}W$. The net result is

$$\frac{d^2W}{dt^2} = \left\{ 4t^2 g(t^2) + \frac{3}{4t^2} \right\} W.$$

The factor of W contains the constant term $4g_1$ and we have reverted to the case with a non-zero constant term. Therefore, this case can be subsumed under the treatment when $g_0 \neq 0$.

(ii) $g_0 = 0$, $g_1 = 0$

In this event both Eq.(6.4.4) and Eq.(6.4.5) are satisfied by $\lambda = 0$. Then Eq.(6.4.6) with $n = 0$ gives

$$\mu^2 - \mu = g_2. \tag{6.4.7}$$

Insertion of Eq.(6.4.7) into Eq.(6.4.6) furnishes

$$n(n + 1 - 2\mu)a_n = g_3 a_{n-1} + \cdots + g_{n+2} a_0 \tag{6.4.8}$$

for $n \geq 1$. The a_n can be determined recursively from Eq.(6.4.8) so long as $2\mu - 1$ is not a positive integer.

Let μ_1 and μ_2 be the two roots of Eq.(6.4.7) with μ_1 chosen so that its real part satisfies $\mathcal{R}(\mu_1) \leq \mathcal{R}(\mu_2)$. Since

$$\mu_1 + \mu_2 = 1$$

$\mathcal{R}(\mu_1) \leq \frac{1}{2}$ so that $2\mu_1 - 1$ cannot be a positive integer. Therefore Eq.(6.4.8) always fixes a_n when $\mu = \mu_1$. Problems arise only for $\mu = \mu_2$. If $\mu_1 = \mu_2$ (which can occur only when both are $\frac{1}{2}$ and $g_2 = -1/4$) both lead to the same a_n so that only one solution of the differential equation is generated. Hence the problem cases are

$$2\mu_2 = k + 1, \ 2\mu_1 = 1 - k \tag{6.4.9}$$

for $k = 0, 1, \ldots$. For the moment these possibilities will be left on one side.

When $\mu = \mu_1$

$$2(1 - \mu_1)a_1 = g_3 a_0$$

from Eq.(6.4.8). Because of the assumed convergence of the expansion for $g(z)$ there is some M such that

$$|g_m| \leq MR^{m-2};$$

there is no loss of generality in taking $M \geq 1$. Since $\mathcal{R}(2\mu_1) < 1$ we see that

$$|a_1| \leq M|a_0|R.$$

Now assume that $|a_m| \leq |a_0|(MR)^m$ for $m = 1, \ldots, n-1$. Then, from Eq.(6.4.8),

$$|a_n| \leq |a_0|(MR)^n / |n + 1 - 2\mu_1| \leq |a_0|(MR)^n \tag{6.4.10}$$

since $|n + 1 - 2\mu_1| \geq 1$. By induction Eq.(6.4.10) is valid without restriction on n.

Consequently, it has been demonstrated that there is a solution of the differential equation

$$w_1(z) = z^{\mu_1} \sum_{m=0}^{\infty} a_m/z^m \qquad (6.4.11)$$

in which the series is uniformly convergent in $|z| > MR$.

When $\mu = \mu_2$ it cannot be asserted that $|n + 1 - 2\mu_2| \geq 1$. However, if

$$M_0 = \sup_n \frac{1}{|n + 1 - 2\mu_2|}$$

the same process reveals that $|a_n| \leq |a_0| \, (MM_0R)^n$. Thus there is a second solution of the differential equation

$$w_2(z) = z^{\mu_2} \sum_{m=0}^{\infty} a_m/z^m \qquad (6.4.12)$$

in which the series is uniformly convergent in $|z| > MM_0R$ other than in the exceptional cases of Eq.(6.4.9).

It may be necessary to impose restrictions on ph z to ensure that w_1 and w_2 are single-valued. Except for any such restriction the properties established justify the substitution of w_1 and w_2 into the differential equation and verification that each is a solution.

In the exceptional cases of Eq.(6.4.9) w_1 remains a solution but a replacement for Eq.(6.4.12) has to be found. Put $w = w_1 W$ in Eq.(6.4.1). The consequent differential equation can be integrated at once and

$$W = B + A \int^z dt/w_1^2(t) \qquad .$$

with A and B arbitrary constants. The formula of Eq.(6.4.11) shows that $w_1(t)/t^{\mu_1}$ is a convergent series for large enough $|t|$ and, moreover, cannot have a zero under the same condition. Accordingly, we can write

$$1/w_1^2(t) = t^{k-1} \sum_{m=0}^{\infty} b_m/t^m$$

by virtue of Eq.(6.4.9). The integration is trivial and, after multiplication by w_1,

$$w = Bw_1(z) + A\{b_k w_1(z) \ln z + z^{\mu_2} \sum_{m=0} c_m/z^m\} \qquad (6.4.13)$$

is the general solution of Eq.(6.4.1) under Eq.(6.4.9). Note that the logarithmic term is absent if $b_k = 0$ and that, if $k = 0$, $c_0 = 0$.

The solution of Eq.(6.4.1) subject to $g_0 = g_1 = 0$ has been resolved completely.

(iii) $g_0 \neq 0$

Here two values of λ, say λ_1 and λ_2 are obtained as solutions of Eq.(6.4.4). The corresponding values μ_1 and μ_2 of μ follow from Eq.(6.4.5). The coefficient of a_{n+1} in Eq.(6.4.6) never vanishes now and all terms in Eq.(6.4.2) can be derived. Unfortunately, the analysis of (ii) cannot be repeated normally to prove that the resulting series is convergent. A notion of why this is so can be attained by supposing that we had succeeded in showing that the right-hand side of Eq.(6.4.6) was bounded by $|a_0| \, n(MR)^n$. In general this bound will be insignificant compared with $n^2 a_n$ when n is large. The conclusion is that a_{n+1} is pretty much the same as $na_n/2\lambda$ for large n. Thus, in general, the most that can be hoped for is that the series is an asymptotic representation of a solution of the differential equation. Only in special circumstances where the right-hand side of Eq.(6.4.6) makes a substantial cancellation in the a_n term is there any chance of securing convergence.

6.5   An error bound for the series

In the case (iii) of the preceding section it was observed that the series stemming from the assumed form of the solution was unlikely to be convergent. Consequently, it is necessary to have an idea of the size of the remainder after n terms have been calculated. It is possible then to decide whether or not the series is asymptotic. The purpose of this section is to confirm that the series does, indeed, provide an asymptotic representation of a solution of the differential equation.

Suppose that the differential equation is

$$\frac{d^2w}{dt^2} = h(t)w$$

where $h(t) = h_0 + h_1/t + \cdots$ for t outside some circle centred on the origin and $h_0 \neq 0$ to fit (iii). Make the substitution $t = x/\sqrt{h_0}$. Then

$$\frac{d^2w}{dz^2} = g(z)w \tag{6.5.1}$$

where $g(z) = h(z/\sqrt{h_0})/h_0$. In view of the assumption on h

$$g(z) = 1 + \frac{g_1}{z} + \cdots \tag{6.5.2}$$

for $|z| > R$, say. It is convenient to discuss the solution of Eq.(6.5.1) subject to Eq.(6.5.2).

In this event Eq.(6.4.4) reduces to $\lambda^2 = 1$. Choose $\lambda_1 = -1$ and $\lambda_2 = 1$. Then, from Eq.(6.4.5), $\mu_1 = -\mu_2 = -g_1/2$. There is no loss of generality in selecting $a_0 = 1$. The a_n resulting from Eq.(6.4.6) with $\lambda = \lambda_1$ and $\mu = \mu_1$ will be denoted by a_{n1}. Similarly a_{n2} is the coefficient when $\lambda = \lambda_2$ and $\mu = \mu_2$.

Now consider the determination of $e_1(z)$ such that

$$w_1(z) = e^{-z} z^{\mu_1} \sum_{m=0}^{n-1} a_{m1}/z^m + e_1(z)$$

is a solution of Eq.(6.5.1). Then

$$\frac{d^2 e_1}{dz^2} = g(z)e_1 - e^{-z} z^{\mu_1} r(n, z) \tag{6.5.3}$$

where the last term originates from the series. Since the satisfaction of Eq.(6.4.6) removes the term in $1/z^{n+2}$ it follows that there is a B_n such that

$$|r(n, z)| \leq B_n / |z|^{n+1} \tag{6.5.4}$$

for $|z| > R$.

Add $-e_1$ to both sides of Eq.(6.5.3) and then convert to an integral equation as in Section 6.2. The aim is to find a solution which is recessive as $z \to +\infty$ so put $e_1(z) = e^{-z} E(z)$ with the result

$$E(z) = \frac{1}{2} \int_z^\infty (1 - e^{2z-2t})[\{g(t) - 1\}E(t) - t^{\mu_1} r(n, t)]dt. \tag{6.5.5}$$

The next step is to try to solve Eq.(6.5.5) by iteration with

$$E_1(z) = -\frac{1}{2} \int_z^\infty (1 - e^{2z-2t})t^{\mu_1} r(n, t)dt \tag{6.5.6}$$

and

$$E_m(z) - E_{m-1}(z) = \frac{1}{2} \int_z^\infty (1 - e^{2z-2t})\{g(t) - 1\}\{E_{m-1}(t) - E_{m-2}(t)\}dt. \tag{6.5.7}$$

When $0 \leq \text{ph}\, z \leq \pi/2$ pick the path of integration to be the same as in Fig. 6.3.2. On it $\mathcal{R}(t - z)$ does not decrease and

$$\left|1 - e^{2z-2t}\right| \leq 2.$$

Furthermore $|t^{\mu_1}| \leq |t|^{\mu_r} \exp(\pi |\mu_i| /2)$ with $\mu_1 = \mu_r + i\mu_i$. It can be arranged that $n > \mu_r$. Hence Eq.(6.5.6) gives

$$|E_1(z)| \leq \frac{B_n \exp(\pi |\mu_i| /2)}{(n - \mu_r) |z|^{n-\mu_r}}.$$

Now there is some B such that $|g(t) - 1| < B/|t|$. Hence induction supplies

$$|E_m(z) - E_{m-1}(z)| \leq \frac{B_n B^{m-1} \exp(\pi |\mu_i| /2)}{(n - \mu_r)^m |z|^{n-\mu_r}}.$$

By increasing n, if necessary, we can be sure that $B < n - \mu_r$. Then the series $\sum_m \{E_m(z) - E_{m-1}(z)\}$ converges absolutely and uniformly. It follows that $|E(z)| = O(1/z^{n-\mu_r})$. Hence, when $0 \leq \mathrm{ph}\, z \leq \pi/2$, there is a solution of Eq.(6.5.1)

$$w_1(z) = e^{-z} z^{\mu_1} \left\{ \sum_{m=0}^{n-1} a_{m1}/z^m + R_1(n, z) \right\} \tag{6.5.8}$$

as $|z| \to \infty$ where

$$R_1(n, z) = O(1/z^n). \tag{6.5.9}$$

When $\pi/2 < \mathrm{ph}\, z \leq \pi$ take the path upwards from z parallel to the imaginary axis until it strikes the circle of radius ρ and then follow the circumference and real axis as in Fig. 6.3.2. As before, the contributions of the circle and real axis vanish as $\rho \to \infty$, leaving an integral of the type

$$\int_0^\infty \frac{dt}{|z + it|^\nu}$$

to be bounded. Since $|z + it|^2 \geq |z|^2 + t^2$ the substitution $t = u^{\frac{1}{2}}$ shows that the integral does not exceed

$$\frac{1}{2} \int_0^\infty \frac{du}{u^{\frac{1}{2}} (|z|^2 + u)^{\frac{1}{2}\nu}} = \frac{(\frac{1}{2}\nu - \frac{3}{2})! \pi^{\frac{1}{2}}}{(\frac{1}{2}\nu - 1)! 2 |z|^{\nu-1}}. \tag{6.5.10}$$

Hence, by putting $\nu = n + 1 - \mu_r$,

$$|E_1(z)| \leq \frac{(\frac{1}{2}n - \frac{1}{2}\mu_r - 1)! \pi^{\frac{1}{2}}}{(\frac{1}{2}n - \frac{1}{2}\mu_r - \frac{1}{2})! 2} \frac{B_n \exp(\pi |\mu_i|)}{|z|^{n-\mu_r}}$$

the change in the factor of $|\mu_i|$ being due to the phase of z lying between $\pi/2$ and π. Moreover

$$|E_m(z) - E_{m-1}(z)| \leq \left\{ \frac{(\frac{1}{2}n - \frac{1}{2}\mu_r - 1)! \pi^{\frac{1}{2}}}{(\frac{1}{2}n - \frac{1}{2}\mu_r - \frac{1}{2})! 2} \right\}^m \frac{B_n B^{m-1} \exp(\pi |\mu_i|)}{|z|^{n-\mu_r}}. \tag{6.5.11}$$

As $\nu \to \infty$ the right-hand side of Eq.(6.5.10) behaves like $(\pi/2\nu)^{\frac{1}{2}}/|z|^{\nu-1}$. Thus, the factor raised to the power m in Eq.(6.5.11) can be made less than 1 by choosing n sufficiently large. Once again the iteration converges and there is no change to Eq.(6.5.8) or to Eq.(6.5.9).

If $-\pi \leq \text{ph}\, z \leq 0$ the same analysis applies except that the paths are drawn in the lower half-plane instead of the upper. Therefore Eq.(6.5.8) and Eq.(6.5.9) hold for $|\text{ph}\, z| \leq \pi$.

Observe that $w_1(z)$, as given by Eq.(6.5.8), is actually independent of n. For, an alteration to n still gives a solution which is recessive as $z \to \infty$ along the real axis and the leading coefficients are unchanged. Since the recessive solution is unique (to within a multiplying constant) the variation in n cannot have had any effect on $w_1(z)$.

The switch from z to $ze^{-\pi i}$ shows that Eq.(6.5.1) has a solution

$$w_2(z) = e^z z^{\mu_2} \left\{ \sum_{m=0}^{n-1} a_{m2}/z^m + R_2(n, z) \right\} \qquad (6.5.12)$$

where

$$R_2(n, z) = O(1/z^n) \qquad (6.5.13)$$

as $|z| \to \infty$ with $0 \leq \text{ph}\, z \leq 2\pi$. Note that $w_2(z)$ is dominant as $z \to \infty$ but recessive as $z \to -\infty$.

The next point to examine is the analytic continuation of w_1 and w_2. The differential equation in Eq.(6.5.1) remains the same if $ze^{-2\pi i}$ is inserted in place of z. Therefore $w_1(ze^{-2\pi i})$ must be a solution of Eq.(6.5.1). It may or may not coincide with $w_1(z)$. But it is certainly not a multiple of $w_2(z)$ being dominant where w_2 is recessive. Hence there are constants A and C_1 such that

$$w_1(z) = Aw_1(ze^{-2\pi i}) + C_1 w_2(z).$$

Let $z \to \infty e^{\pi i}$. From the formulae already established it follows that $A = e^{2\pi i \mu_1}$ and so

$$w_1(z) = e^{2\pi i \mu_1} w_1(ze^{-2\pi i}) + C_1 w_2(z). \qquad (6.5.14)$$

Likewise

$$w_2(z) = e^{-2\pi i \mu_2} w_2(ze^{2\pi i}) + C_2 w_1(z). \qquad (6.5.15)$$

These relations between the analytic continuations are known as *connection formulae*. The question of how to find the constants C_1 and C_2 will be left on one side for the time being.

Notice that, for $\pi \leq \text{ph}\, z \leq 3\pi/2 - \delta$, $w_2(z)$ is exponentially damped according to Eq.(6.5.12) whereas $w_1(ze^{-2\pi i})$ is exponentially increasing by virtue of Eq.(6.5.8). Hence Eq.(6.5.14) implies that $w_1(z)$ continues to have the form of Eq.(6.5.8) for this extended range of ph z. Similar remarks apply when ph z is limited according to $-\pi \geq \text{ph}\, z \geq \delta - 3\pi/2$. The range of Eq.(6.5.12) may be extended in a like manner via Eq.(6.5.15).

All this information is summarised in

Theorem 6.5.1 *There are solutions $w_1(z)$ and $w_2(z)$ of Eq.(6.5.1) which, together with their analytic continuations, can be expressed as*

$$w_1(z) = e^{-z}z^{\mu_1}\left\{\sum_{m=0}^{n-1} a_{m1}/z^m + R_1(n,z)\right\}$$

where $R_1(n,z) = O(1/z^n)$ as $|z| \to \infty$ with $-3\pi/2 + \delta \le \text{ph } z \le 3\pi/2 - \delta$ and

$$w_2(z) = e^{z}z^{\mu_2}\left\{\sum_{m=0}^{n-1} a_{m2}/z^m + R_2(n,z)\right\}$$

where $R_2(n,z) = O(1/z^n)$ as $|z| \to \infty$ with $-\pi/2 + \delta \le \text{ph } z \le 5\pi/2 - \delta$.

6.6　Connection formulae

Here the matter of determining the constants C_1 and C_2 in Eq.(6.5.14) and Eq.(6.5.15) is returned to. Let there be a circle, centre the origin, of radius $\rho > R$. By Cauchy's theorem

$$a_{p1} = \frac{1}{2\pi i}\int_{\rho e^{-\pi i}}^{\rho e^{\pi i}} t^{p-1}\sum_{m=0}^{n-1}\frac{a_{m1}}{t^m}dt$$

so long as $p \le n - 1$. From Theorem 6.5.1

$$a_{p1} = \frac{1}{2\pi i}\int_{\rho e^{-\pi i}}^{\rho e^{\pi i}}\left\{\frac{e^t}{t^{\mu_1}}w_1(t) - R_1(n,t)\right\}t^{p-1}dt.$$

The contour involving R_1 is deformed now into a large circle plus two sides of the negative real axis. There is no contribution as the large circle goes off to infinity because of the bound on R_1. The discontinuity of R_1 across the negative real axis is

$$R_1(n,ze^{\pi i}) - R_1(n,ze^{-\pi i}) = C_1 e^{-z}w_2(ze^{\pi i})/(ze^{\pi i})^{\mu_1}$$

from Eq.(6.5.14) and Theorem 6.5.1. Hence

$$\begin{aligned}a_{p1} = {} & \frac{1}{2\pi i}\int_{\rho e^{-\pi i}}^{\rho e^{\pi i}} t^{p-\mu_1-1}e^t w_1(t)dt \\ & + \frac{(-)^p}{2\pi i}e^{-\pi i\mu_1}\int_{\rho}^{\infty} C_1 e^{-t}t^{p-\mu_1-1}w_2(te^{\pi i})dt.\end{aligned}\tag{6.6.1}$$

The contribution of the first integral in Eq.(6.6.1) is $O(\rho^p)$ since $e^t w_1(t)/t^{\mu_1}$ is bounded on the contour. For the second integral use Theorem 6.5.1 to obtain

$$\begin{aligned}&\int_{\rho}^{\infty} e^{-t}t^{p-\mu_1-1}w_2(te^{\pi i})dt \\ = {} & \int_{\rho}^{\infty} e^{-2t}e^{\pi i\mu_2}t^{p+\mu_2-\mu_1-1}\left\{\sum_{m=0}^{q-1}\frac{a_{m2}}{(te^{\pi i})^m} + R_2(q,te^{\pi i})\right\}dt.\end{aligned}\tag{6.6.2}$$

Now

$$\int_\rho^\infty e^{-t} t^\nu dt = \nu! + O\{\rho^{\nu+1}/(\nu+1)\}$$

and $\mu_2 = -\mu_1$. Provided that $p > \mathcal{R}(2\mu_1) + q$ the right-hand side of Eq.(6.6.2) is

$$e^{\pi i \mu_2} \sum_{m=0}^{q-1} (p + 2\mu_2 - 1 - m)!(-)^m a_{m2}/2^{p+2\mu_2-m}$$
$$+O(\rho^p) + O\{(p + 2\mu_2 - 1 - q)!\}$$

in view of the bound on R_2.

On combining these results we have

$$a_{p1} = \frac{(-)^p}{2\pi i} e^{2\pi i \mu_2} [C_1 \sum_{m=0}^{q-1} (p + 2\mu_2 - 1 - m)!(-)^m a_{m2} 2^{m-p-2\mu_2}$$
$$+O\{(p + 2\mu_2 - 1 - q)!\}] \qquad (6.6.3)$$

when p is large enough. Similarly

$$a_{p2} = -\frac{1}{2\pi i} [C_2 \sum_{m=0}^{q-1} (p + 2\mu_1 - 1 - m)! a_{m1} 2^{m-p-2\mu_1}$$
$$+O\{(p + 2\mu_1 - 1 - q)!\}]. \qquad (6.6.4)$$

Once a sufficient number of the coefficients a_{m1} and a_{m2} has been calculated Eq.(6.6.3) and Eq.(6.6.4) offer approximations to the constants in the connection formulae. In general, it will be necessary for q and p to be quite large to secure a suitable accuracy. For then

$$C_2 = -2\pi i a_{p2}/\{\sum_{m=0}^{q-1} (p + 2\mu_1 - 1 - m)! a_{m1} 2^{m-p-2\mu_1}\} + O(1/p^q) \qquad (6.6.5)$$

which indicates that a minimal requirement is likely to be that p and q should both exceed 10.

Example 6.6.1 To illustrate the above theory a simple example will be considered in which it is easy to evaluate the coefficients. Normally, recourse to symbolic manipulation and numerical technique will be necessary for more complicated differential equations (see Olde Daalhuis & Olver 1995).

The differential equation to be discussed is

$$\frac{d^2 w}{dz^2} = \left\{ 1 + \frac{2\nu}{z} + \frac{\nu(\nu-1)}{z^2} \right\} w.$$

Then $\mu_1 = -\mu_2 = -\nu$ and Eq.(6.4.6) provides

$$a_{n+1,1} = -(n + 2\nu)a_{n1}/2$$

for $n = 0, 1, \ldots$. With $a_{01} = 1$ the solution of the recurrence relation is

$$a_{p1} = \frac{(2\nu + p - 1)!(-)^p}{(2\nu - 1)!2^p}. \tag{6.6.6}$$

On the other hand

$$2(n+1)a_{n+1,2} = n(n+1-2\nu)a_{n2}$$

which shows that $a_{p2} = 0$ for $p \geq 1$.

On account of the form of a_{p2} we can take q as large as we like in Eq.(6.6.3) without affecting the sum in Eq.(6.6.3). Therefore the error term can be dropped and

$$a_{p1} = (p + 2\nu - 1)!\frac{(-)^p e^{2\pi i \nu} C_1}{2\pi i 2^{p+2\nu}}$$

or

$$C_1 = \pi i e^{-2\pi i \nu} 2^{2\nu+1}/(2\nu - 1)!$$

by virtue of Eq.(6.6.6).

It is transparent that $C_2 = 0$.

6.7   Hyperasymptotics

The hyperasymptotic expansion of the remainder $R_1(n, z)$ is the subject to be studied in this section. With z outside the circle of radius ρ introduced in the last section

$$\frac{1}{2\pi i} \int_{\rho e^{-\pi i}}^{\rho e^{\pi i}} \frac{t^{n-1}}{t-z} \sum_{m=0}^{n-1} \frac{a_{m1}}{t^m} dt = 0$$

since there are no singularities inside the circle. Hence

$$\frac{1}{2\pi i} \int_{\rho e^{-\pi i}}^{\rho e^{\pi i}} \left\{ \frac{e^t}{t^{\mu_1}} w_1(t) - R_1(n,t) \right\} \frac{t^{n-1}}{t-z} dt = 0.$$

The contour of the integral containing R_1 is deformed now as in the preceding section. In the process the pole at $t = z$ is captured. There results

$$2\pi i z^{n-1} R_1(n, z) = -\int_{\rho e^{-\pi i}}^{\rho e^{\pi i}} \frac{e^t w_1(t)}{t^{\mu_1}(t-z)} dt$$

$$+ (-)^n e^{-\pi i \mu_1} C_1 \int_{\rho}^{\infty} \frac{t^{n-1-\mu_1}}{t+z} e^{-t} w_2(t e^{\pi i}) dt. \tag{6.7.1}$$

The first integral of Eq.(6.7.1) is estimated easily as $|z| \to \infty$ because $|z|$ will be much larger than ρ. As regards the second integral it is a Stieltjes transform and amenable to the methods described in Sections 5.4 and 5.5. If Theorem 6.5.1 is cited a series of $J_0(\mu, z)$ together with an error term is obtained. Such a version could be convenient for dealing with Stokes' phenomenon.

6.8   Parameter with zero

The treatment of a differential equation with a parameter in Section 6.3 is unsatisfactory if $g(z)$ possesses a zero in the domain of interest. A canonical problem to cope with the presence of a zero is the differential equation

$$\frac{d^2w}{dz^2} = \{k^2 z^m + \psi(z)\}w \qquad (6.8.1)$$

where m is a positive integer and ψ is a regular function. Suppose that $W(z)$ is a solution of

$$\frac{d^2W}{dz^2} = z^m W. \qquad (6.8.2)$$

Then, with $|k|$ large, a first approximation to a solution of Eq.(6.8.1) is $W(\kappa z)$ where $\kappa^{m+2} = k^2$. This suggests trying to solve Eq.(6.8.1) by

$$w = A(z)W(\kappa z).$$

Then

$$2\kappa W'(\kappa z)A'(z) + W(\kappa z)A''(z) = \psi(z)W(\kappa z)A(z).$$

The dominant term as $|k|$ increases is the first one on the left-hand side. Therefore, the differential equation will not be satisfied unless $A'(z) = 0$ i.e. $A(z) =$ constant. But that returns w to the first approximation and precludes any correction for the presence of ψ. Consequently, a more elaborate form must be selected for w.

Try instead

$$w = A(z)W(\kappa z) + \kappa B(z)W'(\kappa z)/k^2 \qquad (6.8.3)$$

in Eq.(6.8.1). The consequent equation is

$$\kappa W'(\kappa z)(2A' + B''/k^2) + W(\kappa z)(A'' + 2z^m B' + mz^{m-1}B)$$
$$= \psi A W(\kappa z) + \psi B\kappa W'(\kappa z)/k^2.$$

Since W and W' are linearly independent their coefficients must vanish. Accordingly

$$2A' = (\psi B - B'')/k^2, \qquad (6.8.4)$$
$$2z^m B' + mz^{m-1}B = \psi A - A''. \qquad (6.8.5)$$

The occurrence of the factor $1/k^2$ in Eq.(6.8.4) floats the notion of introducing the expansions

$$A(z) = 1 + \sum_{p=1} A_p(z)/k^{2p}, \; B(z) = \sum_{p=0} B_p(z)/k^{2p}. \qquad (6.8.6)$$

Equating the coefficients of the powers of $1/k$ and integrating we derive

$$A_{p+1}(z) = -\tfrac{1}{2}B_p'(z) + \int^z \psi(t)B_p(t)dt, \tag{6.8.7}$$

$$B_p(z) = \frac{1}{2z^{m/2}} \int_0^z \frac{1}{t^{m/2}}\{\psi(t)A_p(t) - A_p''(t)\}dt. \tag{6.8.8}$$

B_p is determined by A_p from Eq.(6.8.8) and then A_{p+1} follows from Eq.(6.8.7). Thus, all the terms can be found recursively, starting from $A_0 = 1$. As a consequence Eq.(6.8.7) and Eq.(6.8.8) furnish a formal solution of Eq.(6.8.4) and Eq.(6.8.5) so that Eq.(6.8.6) leads to a w which satisfies formally the differential equation in Eq.(6.8.1).

It is by no means obvious that Eq.(6.8.7) and Eq.(6.8.8) provide functions $A_p(z)$ and $B_p(z)$ which are regular at the origin. B_p would have been certainly singular if the lower limit of integration in Eq.(6.8.8) had been chosen to be other than zero. More can be said. Suppose $\psi(z)$ contains a term whose variation with z is z^n where n is a non-negative integer since ψ is regular. Clearly, $B_0(z)$ will be singular unless $n \geq m-1$. In other words, ψ must have a zero of at least $m-1$ at the origin if our procedure is to succeed.

In fact, $B_0(z)$ is a multiple of z^{n-m+1} so that $A_1(z)$ contains $-(n-m+1)z^{n-m}$ and z^{2n-m+2}. Evidently, $A_1(z)$ is singular also unless $n \geq m-1$. However, $B_1(z)$ is singular when $n = m - 1$ except for $m = 1$. The other cases when $B_1(z)$ is not singular for $n < 2m+1$ are $n = m$, $m+1$ but it turns out that $n = m+1$ is excluded by $B_2(z)$. Continuing step by step we discover that the only permitted form for n is $m + q(m + 2)$ with q a non-negative integer. This restricts $\psi(z)$ to being of the type $C + z^m\phi(z^{m+2})$ with $\phi(z)$ a regular function and C absent if $m \neq 1$. Only subject to this constraint does our procedure solve the differential equation.

Theorem 6.8.1 *The differential equation*

$$\frac{d^2w}{dz^2} = \{k^2 z^m + C + z^m\phi(z^{m+2})\}w$$

can be solved by the procedure described above provided that m is a positive integer, $\phi(z)$ is regular and $C = 0$ if $m \neq 1$.

Differential equations which possess a zero in the factor of k^2 but whose structure is more complicated than z^m have to be transformed into the canonical form of Eq.(6.8.1) before Theorem 6.8.1 can be applied. Often the ensuing ψ is far from simple.

Exercises on Chapter 6

1. Find a solution of

$$\frac{d^2w}{dz^2} = \left(1 + \frac{1}{z^3}\right)w$$

which is recessive as $z \to +\infty$ and bound the error.

2. In Example 6.3.2 show that the error is bounded by $\exp(\pi/8\,|\xi|) - 1$ for $\pi/2 < \mathrm{ph}\,\xi \le \pi$ and by $\exp(\pi/4\,|\mathcal{R}\xi|) - 1$ for $\pi < \mathrm{ph}\,\xi < 3\pi/2$. What are the corresponding statements in terms of z?

3. Obtain WKB approximations for

$$\frac{d^2w}{dx^2} = k^2 \left(1 - \frac{1}{x}\right) w$$

when x is positive and not near 1. Show that, as $x \to \infty$, $e^{kx}/x^{\frac{1}{2}k}$ is an approximation to a solution.

4. Find asymptotic solutions in series for large $|z|$ of

$$\frac{d^2w}{dz^2} = \left(\frac{z+1}{z}\right)^{\frac{1}{2}} w$$

calculating the first three terms. For what ranges of ph z are these series valid?

5. Find connection formulae for the differential equation

$$\frac{d^2w}{dz^2} = \left(1 + \frac{1}{z} - \frac{1}{4z^2}\right) w.$$

6. Investigate the possibility of finding connection formulae for the solutions of the differential equation of Exercise 4.

7. Obtain the formula analogous to Eq.(6.7.1) for $R_2(n, z)$.

8. Show that

$$\frac{d^2w}{dz^2} = \left\{k^2 z^m + \frac{C}{z^2} + z^m \phi(z^{m+2})\right\} w$$

can be solved by the method of Section 6.8, taking W as a known solution of

$$\frac{d^2W}{dz^2} = \left(z^m + \frac{C}{z^2}\right) W.$$

9. In the differential equation

$$\frac{d^2w}{dx^2} = k^2 \left(1 - \frac{1}{x}\right) w$$

change x to ζ where

$$\begin{aligned}
\tfrac{2}{3}\zeta^{3/2} &= (x^2 - x)^{\frac{1}{2}} - \ln\left\{x^{\frac{1}{2}} + (x-1)^{\frac{1}{2}}\right\} & (x \ge 1,\, \zeta \ge 0), \\
\tfrac{2}{3}(-\zeta)^{3/2} &= \cos^{-1} x^{\frac{1}{2}} - (x - x^2)^{\frac{1}{2}} & (0 < x \le 1,\, \zeta \le 0)
\end{aligned}$$

and put $w = (dx/d\zeta)^{\frac{1}{2}} W$. Show that

$$\frac{d^2W}{d\zeta^2} = (k^2\zeta + \psi)W$$

where

$$\psi = \frac{\zeta(3 - 8x)}{16x(x-1)^3} + \frac{5}{16\zeta^2}.$$

Deduce that a first approximation to w is

$$\left(\frac{x\zeta}{x-1}\right)^{1/4} \mathrm{Ai}(k^{2/3}\zeta).$$

Does this bear any resemblance to the approximations of Exercise 3 when x is large and when x is small?

What is the second approximation to w?

Appendix
INTRODUCTION TO NONSTANDARD ANALYSIS

A.1 Basic ideas

Many who deal with the practical applications of mathematics do not bother much about the foundations but are probably aware that conventional mathematics can be developed from set theory. Usually mathematicians will have met an informal discussion of sets at some stage in their training without worrying about the axioms on which they are based. The aim of this appendix is to provide a similar informal discussion of nonstandard analysis. It is not intended to cover all the subtleties of the subject but to provide sufficient information for a working mathematician to be able to use nonstandard analysis whenever it is helpful.

One formal system of axioms for set theory is the Zermelo-Fraenkel system. Without going into the details of the axioms the essence of the system is that, given a set E and a property P, there is always a set F which is part of E and consists of those elements $x \in E$ which have the property P. In 1977 Nelson showed that by adding a prescription for when a set is standard the theory of nonstandard analysis, which had been introduced by Robinson in 1966 via mathematical logic, could be obtained. Thus, every set is either standard or nonstandard. So long as the definition of standard encompasses conventional mathematics there is no change to conventional mathematics but some new (nonstandard) entities become available. In fact, in Nelson's article, there is a theorem which states that any conventional mathematics established by nonstandard analysis could be proved also by Zermelo-Fraenkel theory alone. This does not mean that nonstandard analysis is useless but that it offers an alternative method of attack which may be more powerful in suitable circumstances. Indeed, Bernstein & Robinson (1966) demonstrated that nonstandard analysis could effect the resolution of unsolved questions.

To decide whether or not a set is standard three rules are furnished. As will be seen the rules result in an implicit definition rather than an explicit one so that experience and usage are necessary to gain confidence in making an

141

identification.

The notation to be employed for two frequently occurring sets is ϕ for the empty set and $\mathbf{N}$ for the set of natural integers i.e.

$$\mathbf{N} = \{0, 1, 2, \ldots\}.$$

The set $\mathbf{N}$ can be constructed from ϕ by taking 0 to be ϕ and then

$$1 = \{\phi\}, 2 = \{0, 1\}, 3 = \{0, 1, 2\}, \ldots.$$

Two other notations are $\mathbf{R}$ for real numbers and $\mathbf{C}$ for complex numbers.

The construction of $\mathbf{N}$ might suggest to you that a standard set consists of standard elements but this inference would be false. A standard set can contain nonstandard elements in the same way that a finite set can contain infinite elements. For instance, the set $\{\mathbf{N}, \mathbf{R}\}$ is finite having just 2 elements but neither $\mathbf{N}$ nor $\mathbf{R}$ is finite.

Subsequently, attaching the adjective conventional to a statement or relation will mean that the statement or relation does not contain any reference to the term standard either explicitly or implicitly.

The first rule for identifying standard sets concerns any conventional relation, denoted by $R(x, y)$, between x and y such as $x < y$ or $x \in y$.

Idealisation Rule *The statement that there is an x such that $R(x, y)$ holds for all standard y **is equivalent to** the statement that, for every standard and finite set F, there is an x (which can depend on F) such that $R(x, y)$ holds for all $y \in F$.*

Suppose that $R(x, y)$ signifies that x and y are natural integers such that $x > y$. When F is a finite set of integers there is $n \in \mathbf{N}$ such that $0 \leq y \leq n$. Hence $x = n + 1$ ensures that $R(x, y)$ holds for all $y \in F$. By the Idealisation Rule we deduce that *there is an integer $x \in \mathbf{N}$ such that $x > y$ for all standard integers $y \in \mathbf{N}$.* Consequently, nonstandard integers exist.

An integer x such that $x > y$ for all standard integers y will be called *unlimited*. If x is an unlimited integer so is $x + 1$ since $x + 1 > x > y$ for all standard integers y and, generally, $x + n$ for $n \in \mathbf{N}$ is unlimited. Thus, there is a profusion of unlimited integers. The notion of unlimited can be extended to finite numbers in $\mathbf{R}$ or $\mathbf{C}$ by saying that x is an unlimited number when there is an unlimited integer n such that $|x| \geq n$. Remark that an unlimited number is finite.

The second rule for standard sets is

Standardisation Rule *Let E be a standard set and P a property which the elements of E may possess. Then there is a standard A, a subset of E, whose standard elements are the standard elements of E which possess the property P.*

Note that, in this rule, the property P does not need to be conventional. Note also that the rule refers only to the standard elements of A. There may be nonstandard elements of A which do not possess the property P. Furthermore, there may be nonstandard elements of E which possess the property P but which are absent from A.

The third rule is

Transfer Rule *Let $F(x, a, b, \ldots)$ be a conventional statement relating $x, a, b, \ldots$ in which the parameters $a, b, \ldots$ are standard. If the statement holds for every standard x then it holds for all x.*

In this rule parameter means a quantity capable of free adjustment. For instance, in the statement

$$|y| < 1 + a \text{ for all } y \in E$$

a and E are parameters but not y. The number 1 can be regarded as a constant or as a standard number according to one's preference. If E were standard and the statement held for every positive standard a the Transfer Rule would apply and the statement would be true for every positive a.

The Transfer Rule has an important consequence. If a statement fails for all standard x then the Transfer Rule implies that it fails for all x. Hence, if a statement is true for some x it must be valid for at least one standard x. This principle is used so often that it will be set out separately and referred to as (T) subsequently.

(T) *If there is some x such that the conventional $f(x, a, b, \ldots)$ holds for standard $a, b, \ldots$ then there is a standard x for which $f(x, a, b, \ldots)$ is true.*

When it happens that $f(x, a, b, \ldots)$ defines x uniquely it follows immediately that x is standard. Hence *all quantities defined uniquely in conventional mathematics are standard*. In particular, **N, R, C,** e, π and so on are standard. So is the interval $[a, b]$ if a and b are standard elements of **R**. Similarly, $x = E \cup F$, $x = E \cap F$ imply that $E \cup F$ and $E \cap F$ are standard when both E and F are standard. Likewise, the set of all subsets of E is standard when E is standard.

If E_1 is a subset of E_2 then $x \in E_1$ implies that $x \in E_2$. This is a conventional statement so that, if E_1 and E_2 are standard, and the statement holds for every standard $x \in E_1$ then it is valid for all x by the Transfer Rule. Thus, to verify that E_1 is a subset of E_2 when E_1 and E_2 are standard it is sufficient to confirm that the standard elements of E_1 are in E_2. In particular, two standard sets are equal when they have the same standard elements.

The last sentence shows that the set A of the Standardisation Rule is unique. This has some interesting implications. Let B be the subset of **N** which contains the integers from 0 to the unlimited ω. Now seek the set A of the Standardisation Rule which has the property $x \in B$. A must contain all the standard integers since they are in B because ω is unlimited. But A and $\mathbf{N}$ are standard sets with the same standard elements and so must coincide. Hence, in this case, $A = \mathbf{N}$. Thus, A can be larger than the set which provides the defining property.

On the other hand, if B contained only nonstandard integers $A = \phi$ because both have the same standard elements. In other words the size of A in the Standardisation Rule is unrelated to the size of the set giving the defining property.

Theorem A.1.1 *All elements of a set E are standard if, and only if, E is standard and finite.*

Proof. The statement that a set E is not contained in a standard finite set G is equivalent to saying that, for every standard finite set F, there is an $x \in E$ such that $x \neq y$ for all $y \in F$. By the Idealisation Rule this is equivalent to saying that there is an $x \in E$ such that $x \neq y$ for all standard y. In other words, the statement that E is contained in a standard finite set G is equivalent to saying that all elements of E are standard.

If E is standard and finite choose $G = E$ and the *if* part of the theorem follows.

If, on the other hand, all elements of E are standard E is contained in a standard finite set G. Thus E is finite. Moreover, the set of all subsets of G is finite and standard. Since E is an element of this set it is standard by what has been proved already. The proof of the theorem is complete. ∎

An obvious conclusion from Theorem A.1.1 is that an infinite standard set must contain nonstandard elements.

A function f, defined for all elements of E, is said to be standard when $f(x)$ is standard for all standard $x \in E$. When $x \in E$ is nonstandard a standard f may return standard or nonstandard values. For example, the function $f(n) = n$ for $n \in \mathbf{N}$ returns nonstandard values for nonstandard n whereas the function $f(n) = 1$ for $n \in \mathbf{N}$ always returns a standard value.

If f, g are both standard and $f(x) = g(x)$ for all standard $x \in E$ then the functions are equal by the Transfer Rule. This indicates that once the definition of a standard function has been fixed for the standard elements of E its definition for the nonstandard elements is settled already. For, if f and g differed only on the nonstandard elements of E, their difference would be zero by the first sentence of this paragraph.

The Transfer Rule is called upon frequently in nonstandard analysis. You may be misled by the notation into thinking that x must be a scalar but that is not the intention. For instance, a result may have been proved for every standard x_1, x_2, x_3. By regarding the standard triple (x_1, x_2, x_3) as x the Transfer Rule extends the result to all x_1, x_2, x_3 in the set under consideration.

What the Transfer Rule enables one to do is, on the one hand, to extend results proved for standard x to all x and, on the other hand, infer the existence of a standard x satisfying a formula when some x does.

Perhaps the generality of the Standardisation Rule should be stressed. For example, if you have standard functions which are continuous on an interval, you can form a standard set of functions whose standard elements are continuous standard functions. Results proved for continuous standard functions may carry over to other members of this standard set if transfer is applicable.

A complex number x is said to be *limited* when there is a standard real number y such that $|x| \leq y$. If $|x| < y$ for every standard positive y then x is called *infinitesimal*. The notation $x \simeq y$ will signify that $x - y$ is infinitesimal.

It is obvious that, if x is unlimited, $1/x$ is infinitesimal. Also, if $x \simeq 0$, $xy \simeq 0$ when y is limited. When y is unlimited no conclusion can be drawn in general; for example, with $x > 0$, $y = 1/x^{\frac{1}{2}}$, $1/x$, $1/x^2$ makes the product infinitesimal, limited, unlimited respectively.

Theorem A.1.2 *There is only one standard infinitesimal and that is 0.*

Proof. If $и$ is standard and $|и| < x$ for all standard $x > 0$ the Transfer Rule implies that $|и| < x$ for all $x > 0$. By choosing $x = |и|/2$ we see that $и = 0$ is the only possibility. ∎

Theorem A.1.3 *If $a_n \simeq b_n$ for $n = 1, 2, \ldots, N$ then*

$$\sum_{n=1}^{N}(a_n - b_n)/N \simeq 0.$$

Proof. If ϵ is any positive standard number $|a_n - b_n| < \epsilon$ and

$$\left|\sum_{n=1}^{N}(a_n - b_n)/N\right| < \epsilon$$

which proves the theorem. ∎

Remark that N could be unlimited in Theorem A.1.3. However, if N is limited, multiplication by N does not affect the formula and

$$\sum_{n=1}^{N}(a_n - b_n) \simeq 0 \quad (N \text{ limited}).$$

Other results can be obtained with more assumptions (see Sections A.2 and A.5 later).

Theorem A.1.4 *If x is* **limited** *there is a unique standard number* st(x) *such that* st$(x) \simeq x$.

Proof. Consider firstly that x is a positive real number. Since x is limited there is a standard M such that $x \leq M$. Let A_x be the standard set which contains all the standard y satisfying $0 \leq y \leq x$; A_x exists by the Standardisation Rule. If standard $y \in A_x$ then clearly $y \in [0, M]$. Since A_x and $[0, M]$ are both standard, A_x is contained in $[0, M]$. Consequently, A_x is bounded above. Take st(x) as the upper bound of A_x; it is standard by (T).

Select any standard $\epsilon > 0$. If $\text{st}(x) + \epsilon < x$ then $\text{st}(x) + \epsilon$ is in A_x but greater than its upper bound which is impossible. If $\text{st}(x) - \epsilon > x$ then $\text{st}(x) - \epsilon$ would bound A_x above which is not possible with $\text{st}(x)$ as the upper bound. Hence $|\text{st}(x) - x| < \epsilon$ and $\text{st}(x) \simeq x$.

If there were another standard number x_1 such that $x_1 \simeq x$ then necessarily $\text{st}(x) - x_1 \simeq 0$. Theorem A.1.2 forces $x_1 = \text{st}(x)$ and the theorem has been demonstrated when x is positive.

When x is negative take $\text{st}(x) = -\text{st}(-x)$. If z is the limited complex number $x + iy$

$$\text{st}(z) = \text{st}(x) + i\,\text{st}(y).$$

The proof of the theorem is complete. ∎

There is a natural extension to a limited vector

$$\mathbf{x} = (x_1, x_2, \ldots, x_n)$$

with n standard of

$$\text{st}(\mathbf{x}) = (\text{st}(x_1), \text{st}(x_2), \ldots, \text{st}(x_n)).$$

Observe that, if $\varkappa \simeq 0$ and $\varkappa > 0$, then $\text{st}(\varkappa) = 0$. Thus, from $x > 0$ can be inferred only $\text{st}(x) \geq 0$. It is transparent that

$$\text{st}(xy) = \text{st}(x)\,\text{st}(y)$$

when both x and y are limited.

A.2   Sequences

Suppose that there is a rule which associates with each standard integer n a standard element a_n of a given standard set E. The function $f(n) = a_n$ is standard by the definition of a standard function given earlier. As pointed out previously no other standard function can have the same property and, in addition, there is no choice about the behaviour of f for nonstandard n. Hence the sequence $\{a_n\}$ which takes the constructed standard values for standard n is unique. Such a sequence will be referred to briefly as a *standard sequence* i.e. in a standard sequence $\{a_n\}$ the a_n is a standard element of a standard set E when n is a standard integer.

Theorem A.2.1 *If $\{a_n\}$ is a standard sequence and $\lim_{n\to\infty} a_n = a$ then a is standard.*

Proof. The existence of a implies the statement:

there is an a such that, for every standard $\epsilon > 0$, the set $n \in \mathbf{N}$ with $|a_n - a| \geq \epsilon$ is finite.

Here $\{a_n\}$ and ϵ are standard. Hence, by (T), there must be a standard b with the same property as a. But a limit is unique and so $b = a$. The theorem is proved. ∎

Having established that a is standard replace 'is finite' in the statement quoted in the proof by 'contains $N(\epsilon)$ elements'. Then, by (T), $N(\epsilon)$ can be taken as standard. In other words

for every standard $\epsilon > 0$, there is a standard $N(\epsilon)$ such that $|a_n - a| < \epsilon$ for every $n \geq N(\epsilon)$.

Theorem A.2.2 *If $\{a_n\}$ is a standard sequence, the statements*
  (i) $\lim_{n \to \infty} a_n = a$,
  (ii)   *a is standard and $a_n \simeq a$ for all unlimited $n \in \mathbf{N}$,*
  (iii) $\mathrm{st}(a_n) = a$ *for all unlimited $n \in \mathbf{N}$*
are equivalent.

Proof. By what has just been shown a is standard and $n \geq N(\epsilon)$, when n is unlimited; so $a_n \simeq a$ and (i) implies (ii).

Since a is limited it follows from Theorem A.1.4 that (ii) and (iii) are equivalent.

When (ii) holds choose a definite unlimited N. Then the statement

there is N such that $|a_n - a| < \epsilon$ for $n > N$

is true for every standard $\epsilon > 0$. Treat N as a constant in this statement. The parameters are standard. The Transfer Rule then asserts that it is true for all $\epsilon > 0$. Hence (i) holds and the theorem is proved. ∎

Corollary A.2.2 *If f is standard, the statements*
  (i) $\lim_{x \to 0} f(x) = a$
  (ii)   *a is standard and $f(x) \simeq a$ for all infinitesimal x,*
  (iii) $\mathrm{st}(f(x)) = a$ *for all infinitesimal x*
are equivalent.

Proof. Assume (i) and that standard x_0 is in the domain in which the limit is taken. Then $\{f(x_0/n)\}$ is a standard sequence with limit a whence Theorem A.2.2 makes a standard. Also, by the definition of limit, for every standard $\epsilon > 0$, there is $\delta > 0$ such that $|f(x) - a| < \epsilon$ for $|x| < \delta$. Since f, ϵ, a are standard (T) informs us that the statement holds for a positive standard δ. Thus (ii) is verified and thereafter the proof goes along the lines of Theorem A.2.2. ∎

The theorem can be applied to limits at other standard points by a simple change of variable.

Robinson's Lemma *If $\{a_n\}$ is any sequence then*
(a) if $a_n = 0$ for all standard n, there is an unlimited $\nu \in \mathbf{N}$ such that $a_n = 0$ for all $n \leq \nu$,

*(b) if $a_n \simeq 0$ for all standard n, there is an unlimited $\nu \in \mathbf{N}$ such that $a_n \simeq 0$
for all $n \leq \nu$.*

Proof. (a) Let $m \in \mathbf{N}$ be such that $a_n = 0$ for all $n \leq m$. The set of m contains
all standard m in $\mathbf{N}$ by hypothesis. Hence, by Theorem A.1.1 the set contains
a nonstandard ν.

(b) We cannot proceed as in (a) because $a_n \simeq 0$ is not a conventional state-
ment. However, let $m \in \mathbf{N}$ be such that $|a_n| < 1/m$ for all $n \leq m$. When m is
standard and positive so is $1/m$; accordingly the set of m contains all standard
m in $\mathbf{N}$. Hence the set contains a nonstandard ν such that $|a_n| < 1/\nu$ for all
$n \leq \nu$. Since ν is unlimited the required result follows. ∎

There are some slight variants on the above theorems which are sometimes
useful. Let k be a fixed standard integer and define $b_n = a_{n+k}$ with $\{a_n\}$
standard. The sequence $\{b_n\}$ is standard so that Theorem A.2.2 can be applied.
Thus, altering a standard number of a_n, with n standard, has no effect on
the conclusions. Even if $\{a_n\}$ is not standard the same argument reveals that
Robinson's Lemma is still valid if a standard number of the a_n are not (a) zero,
(b) infinitesimal for standard n.

Another result concerns the persistence of a formula from one unlimited
integer to others.

Corollary A.2.2a *If there is an unlimited integer ω such that, in the standard
sequence $\{a_n\}$, $a_n \simeq 0$ for every unlimited $n \leq \omega$ then $a_n \simeq 0$ for all unlimited
n.*

Proof. Suppose, on the contrary, that there is some $m > \omega$ for which a_m is not
infinitesimal. Then, there is a standard δ such that $|a_m| > \delta$. Define the set of
integers S by $n \in S$ if $n \in \mathbf{N}$ and $|a_n| > \delta$. This defines S uniquely and, since $\mathbf{N}$,
$\{a_n\}$ and δ are standard, makes it standard. S contains an unlimited element
m and so, by virtue of Theorem A.1.1, must be infinite. Moreover, S must
contain some standard n being standard and non-empty. The set $n \in S$ and
$n \leq \omega$ contains all the standard elements of S and hence contains an unlimited
integer ν i.e. there is unlimited $\nu \leq \omega$ such that $|a_\nu| > \delta$. But that contradicts
the hypothesis of the Corollary and so the opening assumption is false. The
proof is finished. ∎

Remark that, if the standard sequence $\{a_n\}$ is such that $a_n \simeq 0$ for all
standard n, then $a_n \simeq 0$ for all unlimited n. For, Robinson's Lemma extends
the formula to some unlimited n and Corollary A.2.2a completes the range.

The validity of Corollary A.2.2a cannot be guaranteed when $\{a_n\}$ is not a
standard sequence. A simple counterexample is $a_n = n\nu_0$ where ν_0 is a fixed
infinitesimal. Here $a_n \simeq 0$ for n up to $1/\nu_0^{\frac{1}{2}}$ say but is certainly not infinitesimal
for $n > 1/\nu_0$. Indeed, the set S in the proof of Corollary A.2.2a is no longer
standard because it has no standard elements but is not empty.

The theorems on sequences can be extended to sequences of functions as will be shown now. It will be assumed that the functions are defined on a standard interval I of the real line.

Theorem A.2.3 *Let f_n be a standard function on I for standard n and f standard. Then, for every $x \in I$, $f_n(x) \to f(x)$ pointwise if, and only if, $f_n(x) \simeq f(x)$ for every standard $x \in I$ and all unlimited integers n.*

Proof. By the Transfer Rule $f_n(x) \to f(x)$ for all standard $x \in I$ if, and only if, $f_n(x) \to f(x)$ for all $x \in I$. For a given standard x $\{f_n(x)\}$ is a standard sequence and $f(x)$ is standard. From Theorem A.2.2 it follows that $f_n(x) \simeq f(x)$ for unlimited n provides the same information and the proof is complete. ∎

Theorem A.2.4 *Under the same conditions as Theorem A.2.3 $f_n \to f$ uniformly on I if, and only if, $f_n(x) \simeq f(x)$ for all $x \in I$ and all unlimited integers n.*

Proof. If $g(x) \simeq 0$ for all $x \in I$, $|g(x)| < \epsilon$ for standard $\epsilon > 0$. Hence g is bounded and $\sup |g(x)| \leq \epsilon$. Hence $\sup |g(x)| \simeq 0$.

Consequently, the statement $f_n(x) \simeq f(x)$ for all $x \in I$ is equivalent to $\sup |f_n(x) - f(x)| \simeq 0$. The sequence $\{\sup |f_n(x) - f(x)|\}$ is standard by (T) and so Theorem A.2.2 applies. The proof is finished. ∎

Robinson's Continuous Lemma *If $f(x) \simeq 0$ for all limited x there is an unlimited X such that $f(x) \simeq 0$ for $|x| \leq X$.*

Proof. Let $m \in \mathbf{N}$ be such that $|f(x)| < 1/m$ for $|x| \leq m$. The set of m includes all the standard m by hypothesis. Therefore, the set contains an unlimited integer ν such that $|f(x)| < 1/\nu$ for $|x| \leq v$. The result stated can be inferred immediately. ∎

A.3   Continuity

Definition A.3.1 *The function f is said to be **S-continuous** at x if, and only if, $f(y) \simeq f(x)$ for all $y \simeq x$.*

Theorem A.3.1 *If f is standard f is S-continuous at standard x if, and only if, f is continuous at x.*

Proof. Given S-continuity it is valid to make the statement:

> there is a $\delta > 0$ such that $|f(x) - f(y)| < \epsilon$ for all y in $|y - x| < \delta$ and for every standard $\epsilon > 0$

by choosing δ as infinitesimal. Keep δ constant. Then the Transfer Rule can be applied legitimately and the statement holds for all $\epsilon > 0$. Thus f is continuous at x.

Conversely, when f is continuous at x, choose any standard $\epsilon > 0$. Then the statement

> there is a $\delta(\epsilon) > 0$ such that $|f(x) - f(y)| < \epsilon$ for all y in $|y - x| < \delta(\epsilon)$

is true. Since the parameters f and ϵ are standard (T) confirms that the statement is true for some standard positive $\delta(\epsilon)$. Now, if $y \simeq x$, $|y - x| < \delta(\epsilon)$ for standard positive $\delta(\epsilon)$ and so $|f(x) - f(y)| < \epsilon$. Since ϵ is arbitrary, $f(x) \simeq f(y)$ and S-continuity has been affirmed. ∎

On account of Theorem A.3.1, the set of x of a standard interval I at which the standard f is continuous is the same as the set of the Standardisation Rule with the property that f is S-continuous at x. Both sets are standard with the same standard elements. As a consequence the statement that the standard f is continuous on the standard I offers the same information as saying that f is S-continuous at every standard x of I. To put it another way, if the standard set of functions which are S-continuous at x is formed by the Standardisation Rule it contains precisely the functions which are continuous at x. In this case there is no ambiguity about the elements of the standard set provided by the Standardisation Rule although generally there is information about the standard elements solely.

When x is limited $\mathrm{st}(x)$ exists and $\mathrm{st}(x) \simeq x$ by Theorem A.1.4. Therefore, if f is standard and S-continuous at x,

$$f(x) \simeq f(\mathrm{st}(x)).$$

The right-hand side is standard and, therefore, limited. Hence $f(x)$ is limited and so $\mathrm{st}(f(x))$ exists. Applying Theorem A.1.4 we conclude that

$$\mathrm{st}(f(x)) = f(\mathrm{st}(x)) \qquad (A.3.1)$$

when f is standard and S-continuous at limited x.

Violation of the conditions of Theorem A.3.1 can destroy any connection between continuity and S-continuity. We illustrate with a few examples.

The function $f(x) = x^2$ is standard. It is continuous at every x because $|f(y) - f(x)| < \epsilon$ for $|y - x| < \delta$ so long as $\delta < (x^2 + \epsilon)^{\frac{1}{2}} - x$. However, f is not S-continuous at unlimited x. For, with $y = x + 1/x$ so that $y \simeq x$,

$$f(y) - f(x) = 2 + 1/x^2$$

which is not infinitesimal.

As an example of a nonstandard f take $f(x) = \varkappa/(\varkappa^2 + x^2)$ with $\varkappa$ a positive infinitesimal. At the standard $x = 0$, f is continuous but not S-continuous because $f(0) - f(\varkappa) = 1/2\varkappa$ which is unlimited.

In contrast, the selection $f(x) = \varkappa \, \mathrm{sgn}\, x$ makes f discontinuous at $x = 0$ but it is S-continuous at every x because $|f(x) - f(y)| \leq 2\varkappa$.

A standard interval I of the real line is said to be *compact* when every $x \in I$ is limited and $\mathrm{st}(x) \in I$.

Theorem A.3.2 *If f is continuous on the compact standard $I = [a, b]$ with $f(a) > 0$ and $f(b) < 0$ then there is $c \in I$ where $f(c) = 0$.*
Proof. Assume that f is standard. Let J be a finite subset of I which contains all standard $x \in I$. Let x_0 be the largest element of J for which $f(x_0) \geq 0$. Since b is standard it is in J and so $x_0 < b$. Also x_0 is limited by assumption and

$$f(\text{st}(x_0)) \geq 0$$

by Eq.(A.3.1). Let x_1 be the next element of J above x_0; then $f(x_1) < 0$. If $x_1 = x_0 + \delta$ and δ is not infinitesimal

$$2x_0 < \text{st}(2x_0 + \delta) \leq \text{st}(2x_1 - \delta) < 2x_1.$$

This means that $\text{st}(x_0 + x_1)/2$ is a point of J between x_0 and x_1 contrary to the definitions of x_0 and x_1. Hence δ must be infinitesimal. By Theorem A.1.4

$$0 \leq f(\text{st}(x_0)) \simeq f(x_1) < 0$$

which forces $f(\text{st}(x_0)) = 0$. Thus the c of the theorem is standard when f is standard.

The Transfer Rule extends the theorem to more general f. ∎

Of course, the theorem does not assert that there is only one zero with $a < c < b$. There can be many others and they may be nonstandard even when f is standard.

A.4   The derivative

The function f is said to be *S-differentiable* at the point a when there exists a standard d such that
$$\frac{f(x) - f(a)}{x - a} = d$$
for all $x \simeq a$, and d is called the *S-derivative* of f.
Theorem A.4.1 *If f is standard, f is S-differentiable at standard a if, and only if, f is differentiable at a.*
Proof. When f is differentiable define

$$g(x) = \frac{f(x) - f(a)}{x - a}.$$

Then, with $g(a) = \lim_{x \to a} g(x)$, g is a continuous standard function and also $g(a) = f'(a)$ with $f'(a)$ standard. By Theorem A.3.1, when $x \simeq a$, $g(x) \simeq g(a)$ and f is S-differentiable; the S-derivative coincides with the usual derivative.

When f is S-differentiable $g(x) \simeq d$ for $x \simeq a$. Define $g(a) = d$ so that g is standard and S-continuous at a. Theorem A.3.1 shows that g is continuous so that f is differentiable at a with $f'(a) = d$. ∎

Although the requirement that d be standard in S-differentiability looks rather restrictive it is not as confining as appears at first sight. Suppose that f is S-differentiable at every standard $x \in I$ with I standard. Then there is a standard function which reproduces the values of d at the standard x and this function may be taken as the derivative for nonstandard x. From another point of view, the standard set of the Standardisation Rule with the property f is S-differentiable is the same as the standard set with f differentiable. In particular, if f is standard, the set of points at which f is differentiable is standard.

For example, let $f(x) = x^2$. If и is any infinitesimal

$$f(x + и) - f(x) = 2иx + и^2$$

which shows that the S-derivative is $2x$ at standard x. The domain $x \geq 0$ contains all standard $x \geq 0$ and so $2x$ can be taken as the derivative throughout $x \geq 0$.

These remarks indicate that there is little point in distinguishing between the notation for the S-derivative and the conventional derivative.

Theorem A.4.2 *If f is standard on the standard I and has derivative $f'(x)$ at x then there is $\delta > 0$ such that*

$$\frac{f(y) - f(x)}{y - x} \simeq f'(x)$$

for $0 < |y - x| < \delta$.

Note that there is no necessity for x to be standard in this theorem.

Proof. We know that the set of points at which f is differentiable is standard—call it J. J is not empty because it contains x. If there were no standard point in J then J would be empty, the empty set being standard and having no standard points. Therefore there is a standard point, say a, in J. Pick any infinitesimal и > 0. Then, on account of the differentiability at a, for $0 < |y - a| < $ и

$$\frac{f(y) - f(a)}{y - a} \simeq f'(a)$$

whence

$$\left| \frac{f(y) - f(a)}{y - a} - f'(a) \right| < \epsilon$$

for all standard $\epsilon > 0$. This is a statement in which f, f', ϵ and a are standard. Therefore, by the Transfer Rule, it holds for all points of J. The proof is terminated. ∎

It is fairly obvious now that all the usual rules for the derivative are valid e.g. $(fg)' = f'g + fg'$, Rolle's theorem, $f' = 0$ on I entails $f = $ constant, $f' > 0$ means f is increasing and we shall not bother to prove them in detail.

For later purposes it is helpful to have a stronger version of differentiability, namely

> The standard f is said to be *strongly differentiable* at the standard a if there is a standard d such that
> $$\frac{f(x) - f(y)}{x - y} \simeq d$$
> for all $x \simeq a$ and all $y \simeq a$ with $x \neq y$.

Observe that, by selecting $y = a$, strong differentiability enforces differentiability with $d = f'(a)$.

Theorem A.4.3 *The standard f is strongly differentiable at all interior points of the standard I if, and only if, f is differentiable and f' is continuous on I.*
Proof. Let $a \in I$ be a standard interior point and $x \simeq a$. Since strong differentiability entails differentiability Theorem A.4.2 gives

$$\frac{f(y) - f(x)}{y - x} \simeq f'(x)$$

for $y \simeq x$ and $y \neq x$. By virtue of strong differentiability at a

$$\frac{f(y) - f(x)}{y - x} \simeq f'(a).$$

Hence $f'(x) \simeq f'(a)$ for $x \simeq a$ and continuity follows from Theorem A.3.1 since f' is standard. The *only if* part of the theorem has been proved.

To show the converse take $x \simeq a$, $y \simeq a$ with $x \neq y$. Differentiability provides, via Rolle's theorem, a c between x and y such that

$$\frac{f(y) - f(x)}{y - x} = f'(c).$$

From the continuity of f' and $c \simeq a$, Theorem A.3.1 reveals that $f'(c) \simeq f'(a)$ whence strong differentiability can be affirmed. The proof is complete. ■

Higher derivatives (when they exist) can be calculated in a natural way since f' is standard when f is. Thus Taylor's expansion

$$f(x) = f(a) + f'(a)(x - a) + \ldots + f^{(n)}(a)(x - a)^n/n! + R_n$$

can be available. When f is infinitely differentiable at a f is said to be *analytic* at a when the remainder R_n satisfies $R_n \simeq 0$ with $|x - a| < \delta$ for some standard δ and n unlimited.

The extension to functions of a complex variable z is immediate. Analytic functions can be defined in the usual way and have the usual properties.

A.5   Integration

As in conventional theory we start with integrals on the real line. Let f be bounded on the interval $[a, b]$ with $a < b$. Given any infinitesimal $\varkappa$ let N be the largest integer such that $a + N\varkappa \leq b$.

Definition A.5.1 $\int_a^b f(x)dx = \mathrm{st}\left\{\sum_{j=0}^N f(t_j)\varkappa\right\}$ *for any infinitesimal $\varkappa$ and every $t_j \in [a + j\varkappa, a + (j+1)\varkappa]$.*

It is necessary to check that the right-hand side has a meaning. Since f is bounded there is M such that $|f| \leq M$ throughout the interval. Hence

$$\left|\sum_{j=0}^N f(t_j)\varkappa\right| \leq M(N+1)\varkappa \leq M(b - a + \varkappa).$$

If f is standard, M can be taken as standard. Therefore, the right-hand side of the inequality is limited for $b - a$ limited. Thereby, the left-hand side is limited and it is permissible to take the standard part of the definition.

It has been demonstrated that the integral certainly exists when a, b and f are standard. Consequently, it can be extended to other (a, b, f) by identifying it with the standard function which takes the standard value of the Definition for the standard triples (a, b, f) with f bounded on $[a, b]$.

Remark The confirmation that a meaning can be attached to the sum in Definition A.5.1 is unaffected by changing f to $|f|$. So $\int_a^b |f(x)|\,dx$ exists and

$$\left|\int_a^b f(x)dx\right| \leq \int_a^b |f(x)|\,dx.$$

If c satisfies $a < c < b$ the additive property of the integral

$$\int_a^b f(x)dx = \int_a^c f(x)dx + \int_c^b f(x)dx$$

can be deduced in a trivial manner.

It is possible that, for some functions, the value of the sum depends upon the choice of the infinitesimal $\varkappa$. The next theorem reveals that this surmise is false for continuous functions.

Theorem A.5.1 *If f is continuous on the compact interval $[a.b]$ the value of the integral is independent of $\varkappa$.*

Proof. It is sufficient to prove the result when f takes real values. Let x, y be standard points of the interval with $x < y$. Assume to begin with that f is standard. The continuity of f on $[x, y]$ ensures that

$$\min f \leq f \leq \max f$$

there. Accordingly, as in the justification of the definition,

$$(y - x)\min f \leq \int_x^y f(t)dt \leq (y - x)\max f.$$

Since the integral has been identified with a standard function the inequality can be extended by the Transfer Rule from the standard triples (y, x, f) to any triples (y, x, f) with f continuous on $[x, y]$. Then, if $F(x) = \int_a^x f(t)dt$,

$$\min f \leq \frac{F(y) - F(x)}{y - x} \leq \max f$$

by the additive property. Pick a standard c and choose $y \simeq c$, $x \simeq c$ with $y \neq x$. Because f is continuous $\max f$ and $\min f$ differ infinitesimally from $f(c)$ on $[x, y]$. Hence $F(x)$ is strongly differentiable at any standard c, and has derivative $f(c)$. Since the integral is a standard function the Transfer Rule extends this result to every point of the interval.

Suppose now that a different infinitesimal is used in the definition of the integral, leading to $F_1(x)$ say. Then $F_1(x)$ is strongly differentiable with derivative $f(c)$ at c. Hence $F'(x) = F_1'(x)$ whence $F(x) - F_1(x) = \text{constant} = 0$ because both integrals vanish at $x = a$. The Transfer Rule extends the theorem to any continuous f. The proof is terminated. ∎

Piecewise continuous functions can be treated by first splitting the integral into portions where there is continuity by the additive property and then applying Theorem A.5.1 to each of the separate portions.

Since the conventional indefinite integral enjoys the property that its derivative reproduces f where f is continuous our integral agrees with the conventional whenever the two exist. The additive property extends the equality to piecewise continuous functions.

Another useful result is

Theorem A.5.2 *If $f(x) \simeq 0$ for $a \leq x \leq b$ with $b - a$ limited*

$$\int_a^b f(x)dx \simeq 0.$$

Proof. Obviously

$$\left| \int_a^b f(x)dx \right| \leq (b - a) \sup |f|.$$

As in Theorem A.2.4 $f(x) \simeq 0$ implies $\sup |f| \simeq 0$. The stated value of the integral can be inferred at once from $b - a$ being limited and the proof is complete. ∎

Corollary A.5.2 *If f is bounded and integrable over $[a - \varkappa_0, b + \varkappa_0]$ with $\varkappa_0$ infinitesimal, then*

$$\int_a^b |f(x) - f(x + \varkappa_0)| \, dx \simeq 0.$$

Proof. Since both $\int_a^b f(x)dx$ and $\int_a^b f(x + \varkappa_0)dx$ exist by hypothesis so does the given integral by the Remark after Definition A.5.1. We can assume $\varkappa_0 > 0$ without loss. Choose any infinitesimal $\varkappa > \varkappa_0$. Then

$$\int_a^b |f(x) - f(x + \varkappa_0)| \, dx \simeq \sum_{j=0}^N |f(a + j\varkappa) - f(a + j\varkappa + \varkappa_0)| \, \varkappa.$$

Also

$$\sum_{j=0}^{N} f(t_j)\varkappa \simeq \int_a^b f(x)dx \simeq \sum_{j=0}^{N} f(t_j')\varkappa$$

whence

$$\sum_{j=0}^{N}\{f(t_j) - f(t_j')\}\varkappa \simeq 0.$$

If $f(a+j\varkappa) \geq f(a+j\varkappa+\varkappa_0)$ choose $t_j = a+j\varkappa$, $t_j' = a+j\varkappa+\varkappa_0$; otherwise reverse the roles of t_j and t_j'. The stated result follows immediately. ∎

Singular integrals occur whenever f is unbounded in the range of integration or the range is infinite. It is customary to define such integrals (when they exist) by means of limits e.g.

$$\int_a^\infty f(x)dx = \lim_{t\to\infty} \int_a^t f(x)dx.$$

Theorem A.5.3 *Let $l_n = \int_a^n f(x)dx$ with $n \in$ **N**. If $\{l_n\}$ is a standard sequence the infinite integral exists if, and only if, $l_n \simeq l$ with l standard for all unlimited n. When the integral exists $\int_a^\infty f(x)dx = l$ and $\int_\omega^\infty f(x)dx \simeq 0$ for any unlimited integer ω.*

Proof. Application of Theorem A.2.2 covers all the assertions except the last which is a consequence of the first part and the additive property. There is nothing more to prove. ∎

There are similar theorems for integrals to $-\infty$ and from $-\infty$ to ∞ but details will be omitted.

For $f(x)$ unbounded at a point, which may be taken as the origin without loss of generality, one is concerned with limits such as

$$\lim_{\epsilon \to +0} \int_\epsilon^b f(x)dx$$

or

$$\lim_{\epsilon \to +0} \left\{ \int_\epsilon^b f(x)dx + \int_{-b}^{-\epsilon} f(x)dx \right\}.$$

Here the theorem analogous to Theorem A.5.3 depends on sequences like $\left\{ \int_{1/n}^b f(x)dx \right\}$ but details are left to the reader.

For many applications it is useful to have

Theorem A.5.4 *If $f(x) \simeq 0$ for all limited $x \geq 0$ and $|f(x)| \leq h(x)$ for all $x \in$ **R** with $x \geq 0$ where standard h is integrable from 0 to ∞ then*

$$\int_0^\infty f(x)dx \simeq 0.$$

Proof. If n is a positive limited integer

$$\int_0^n f(x)dx \simeq 0$$

by virtue of Theorem A.5.2. Robinson's Lemma assures us that there is an unlimited integer ω such that

$$\int_0^\omega f(x)dx \simeq 0.$$

Also

$$\left| \int_\omega^\infty f(x)dx \right| \leq \int_\omega^\infty h(x)dx \simeq 0$$

from Theorem A.5.2. Addition completes the proof. ∎

Actually $f(x) \simeq 0$ for more x than specified in Theorem A.5.4 by Robinson's Continuous Lemma but that fact is not needed in the proof.

Corollary A.5.4 *If $\int_{-\varkappa_0}^\infty |f(x)|\, dx$ exists then*

$$\int_0^\infty |f(x) - f(x + \varkappa_0)|\, dx \simeq 0$$

for infinitesimal $\varkappa_0$.

Proof. The proof is the same as for Theorem A.5.4 but calling on Corollary A.5.2. ∎

By replacing f by $f - g$ in Theorem A.5.4 we obtain

Theorem A.5.5 *If $f(x) \simeq g(x)$ for all limited $x \geq 0$ and $|f(x)| \leq h(x)$, $|g(x)| \leq h(x)$ for all $x \in \mathbf{R}$ with $x \geq 0$ where standard h is integrable from 0 to ∞ then*

$$\int_0^\infty f(x)dx \simeq \int_0^\infty g(x)dx.$$

A variant of this theorem is

Theorem A.5.6 *Let $f(x) = g(x)h(x)$ with $g(x) \simeq 0$ for all limited x and $h > 0$ integrable from 0 to ∞. If $|f(x)| \leq 2h(x)$ for all $x \in \mathbf{R}$*

$$\int_0^\infty f(x)dx = \varkappa \int_0^\infty h(x)dx$$

for some infinitesimal $\varkappa$.

Proof. For limited positive $n \in \mathbf{N}$

$$\left| \int_0^n f(x)dx \right| \leq \sup |g| \int_0^n h(x)dx$$

so that

$$\int_0^n f(x)dx \Big/ \int_0^n h(x)dx \simeq 0.$$

By Robinson's Lemma this is true for an unlimited ω i.e.

$$\int_0^\omega f(x)dx = \varkappa \int_0^\omega h(x)dx.$$

Since $\int_\omega^\infty f(x)dx \simeq 0$ and $\int_\omega^\infty h(x)dx \simeq 0$ as in theorem A.5.4 the stated formula follows. ∎

Corresponding theorems can be developed for infinite series by employing the sequence $\{s_n\}$ where $s_n = \sum_{m=0}^{n} a_m$. They will be stated without proof.

Theorem A.5.7 *If $\{s_n\}$ is a standard sequence $\sum_{n=0}^{\infty} a_n$ exists if, and only if, $s_n \simeq l$ with l standard for all unlimited n. When the infinite series exists $\sum_{n=0}^{\infty} a_n = l$ and $\sum_{n=\omega}^{\infty} a_n \simeq 0$ for any unlimited ω.*

Theorem A.5.8 *If $a_n \simeq 0$ for all limited integers n and $|a_n| \leq h_n$ for every $n \in \mathbf{N}$ where $\sum_{n=0}^{\infty} h_n$ is convergent and h_n is standard, then $\sum_{n=0}^{\infty} a_n \simeq 0$.*

Theorem A.5.9 *If $a_n \simeq b_n$ for all limited integers n and $|a_n| \leq h_n$, $|b_n| \leq h_n$ where $\sum_{n=0}^{\infty} h_n$ is convergent and h_n standard, then $\sum_{n=0}^{\infty} a_n \simeq \sum_{n=0}^{\infty} b_n$.*

Theorem A.5.10 *If $a_n = b_n h_n$ with $b_n \simeq 0$ for all limited integers n, standard $h_n > 0$ with $\sum_{n=0}^{\infty} h_n$ convergent and $|a_n| \leq 2h_n$ for all $n \in \mathbf{N}$ then*

$$\sum_{n=0}^{\infty} a_n = \varkappa \sum_{n=0}^{\infty} h_n$$

for some infinitesimal $\varkappa$.

The Maclaurin–Cauchy test for convergence offers a connection between series and integrals.

Theorem A.5.11 *If $f(x) > 0$ and f decreases steadily then*

$$\sum_{m=0}^{n} f(m) \geq \int_{0}^{n+1} f(x)dx \geq \sum_{m=1}^{n+1} f(m)$$

for any $n \in \mathbf{N}$. If, in particular, $f(x) \simeq 0$

$$\sum_{m=0}^{n} f(m) \simeq \int_{0}^{n} f(x)dx.$$

Proof. For any integer m the decrease in f enforces

$$f(m) \geq \int_{m}^{m+1} f(x)dx \geq f(m+1).$$

Addition over m supplies the first part of the theorem. When $f(x) \simeq 0$ the two extremes of the inequality differ only by an infinitesimal. The second part is an instant consequence. ∎

References

Abramowitz, M. and Stegun, I. A. 1965 *Handbook of Mathematical Functions.* Dover, New York.

Bernstein, A. R. and Robinson, A. 1966 *Pacific Journal of Math.* **16**(3), 421–431.

Berry, M. V. and Howls, C. J. 1990 *Proc. Roy. Soc. Lond.* A**430**, 653–668.

Berry, M. V. and Howls, C. J. 1991 *Proc. Roy. Soc. Lond.* A**434**, 657–675.

Berry, M. V. 1991 *Proc. Roy. Soc. Lond.* A**435**, 437–444.

Bleistein, N. and Handelsman, R. A. 1975 *Asymptotic Expansions of Integrals.* Holt, Rinehart and Winston, New York.

Boyd, W. G. C. 1990 *Proc. Roy. Soc. Lond.* A**429**, 227–246.

Boyd, W. G. C. 1993 *Proc. Roy. Soc. Lond.* A**440**, 493–518.

Boyd, W. G. C. 1994 *Proc. Roy. Soc. Lond.* A**447**, 609–630.

Chester, C., Friedman, B. and Ursell, F. 1957 *Proc. Cambridge Philos. Soc.* **53**, 599–611.

Dingle, R. B. 1973 *Asymptotic Expansions: Their Derivation and Interpretation.* Academic Press, London.

Howls, C. J. 1992 *Proc. Roy. Soc. Lond.* A**439**, 373–396.

Izumi, S. 1927 *Japan J. Math.* **4**, 29–32.

Jones, D. S. 1982 *The Theory of Generalised Functions.* Cambridge University Press, Cambridge.

Lighthill, M. J. 1958 *Fourier Analysis and Generalised Functions.* Cambridge University Press, Cambridge.

McLeod, J. B. 1992 *Proc. Roy. Soc. Lond.* A**437**, 343–354.

Nelson, E. 1977 *Bull. Amer. Math. Soc.* **83**, 1165–1198.

Olde Daalhuis, A. B. 1992 *I.M.A. Jour. Appl. Math.* **49**, 203–216.

Olde Daalhuis, A. B. 1993 *Proc. Roy. Soc. Edin.* A**123**, 731–743.

Olde Daalhuis, A. B. and Olver, F. W. 1994 *Proc. Roy. Soc. Lond.* A**445**, 39–56.

Olde Daalhuis, A. B. and Olver, F. W. J. 1995 *Methods and Applications of Analysis.* **2**, 348–367.

Olver, F. W. J. 1974 *Asymptotics and Special Functions.* Academic Press, New York.

160

Olver, F. W. J. 1991a *SIAM J. Math. Anal.* **22**, 1460–1474.

Olver, F. W. J. 1991b *SIAM J. Math. Anal.* **22**, 1475–1489.

Olver, F. W. J. 1994 *Methods Applic. Analysis.* **1**, 1–13.

Paris, R. B. 1992a *J. Comp. appl. Math.* **41**, 117–133.

Paris, R. B. 1992b *Proc. Roy. Soc. Lond.* **A436**, 165–186.

Paris, R. B. and Wood, A. D. 1992 *J. Comp. appl. Math.* **41**, 135–143.

Paris, R. B. and Wood, A. D. 1995 *Bull. IMA.* **31.** 21–28.

Robert, A. 1988 *Nonstandard Analysis.* Wiley, New York.

Robinson, A. 1966 *Non Standard Analysis.* North-Holland, Amsterdam.

Stieltjes, T. J. 1886 *Ann. Sci. École Norm. Sup.* [3], **3**, 201–258.

van den Berg, I. 1987 *Nonstandard Asymptotic Analysis.* Springer-Verlag, Berlin.

Wong, R. 1989 *Asymptotic Approximations of Integrals.* Academic Press, Boston.

Index